T0076232

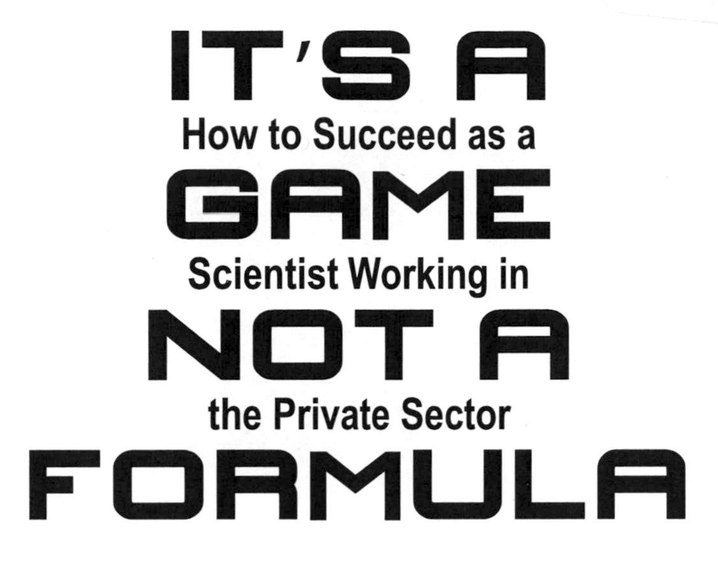
IT'S A
How to Succeed as a
GAME
Scientist Working in
NOT A
the Private Sector
FORMULA

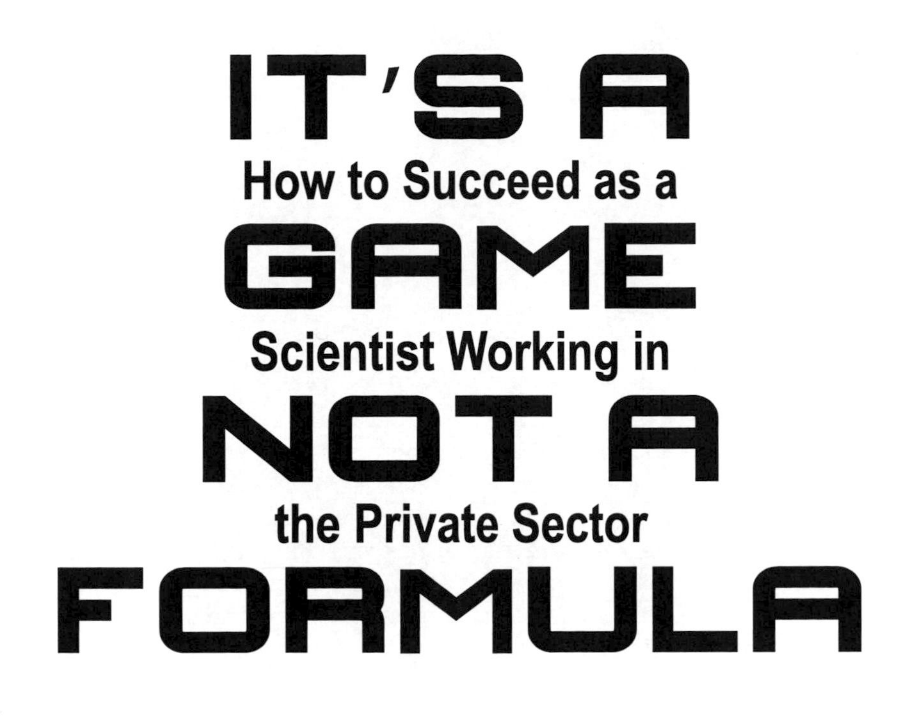

DAVID M. GILTNER

SPIE PRESS
Bellingham, Washington USA

Library of Congress Control Number: 2021944565

Published by
SPIE
P.O. Box 10
Bellingham, Washington 98227-0010 USA
Phone: +1 360.676.3290
Fax: +1 360.647.1445
Email: books@spie.org
Web: www.spie.org

Printed in the United States of America.
First Printing.
For updates to this book, visit http://spie.org and type "PM332" in the search field.

Table of Contents

Preface

Back in 2010, I published a book called *Turning Science into Things People Need.* It contained interviews with ten different scientists who had built successful careers in the private sector. I wrote the book during an extended job transition as something to do to fill the gap while I was looking for just the right opportunity to fit the new direction that I had decided to take my career. I didn't know for sure what I was going to do with the book once it was published. I had thought for some time that I'd like to write a book so that I could travel on a speaking circuit, and so when I met a publisher who was helping people create books based on interviews around a topic that excited them, I decided that the time was right.

The next task was to decide what I would interview people about as the core topic of my book. Since I was in the middle of a job search at the time, I had been reminded of the challenges a scientist faces when they decide to forgo the presumed 'conventional' career path of a professor and build their career path in the private sector instead. After a few weeks of consideration, I chose the subtitle 'Voices of Scientists Working in Industry.' By telling the stories of other successful scientists, I hoped I could attempt to open other young scientists' eyes to new career options and how to sell their unique scientist skills in the private sector. I knew that when I was nearing the end of my postgraduate studies, I would have appreciated a book that told the stories of other scientists who had built private-sector careers, and I had yet to find any other book that had addressed this topic. At the time, I was about a dozen years into my own career, and I knew many other scientists who had also build successful industry careers. They would give me a great pool from which to select interview candidates.

I had no idea where the journey that began with that book would take me. The seminar I put together to begin my speaking career focused on two key topics pulled from the book interviews: (1) jobs that scientists will enjoy and be good at in industry, and (2) the unique strengths that a scientist brings to the private sector. As I began traveling and speaking at universities, I learned so much more about where a typical university science education fails to prepare us for private-sector careers. When I left my job in early 2017 to found TurningScience, I spent time interviewing industry managers to learn their perspective on how well early-career scientists made the transition into industry. From those conversations, I realized that many of the things we scientists were taught in academia actually work against us when we transition into the private sector.

I realized that we scientists tend to look for the 'right way' to do things. We are used to chasing after formulas and theories that describe the behavior of the universe, and this expectation that there is a 'right answer' tends to permeate much of what we do. But the world outside the controlled environment of the science lab doesn't work that way. It occurred to me that the colleagues of mine who were most successful approached their work as if it were a game, with rules that need to be followed but with no clear 'right way' to do things. They took risks, made decisions quickly, and didn't overthink things—fundamentally different from what the early-career scientists I'd worked with so often seemed to do.

I realized that what early-career scientists need even more than lectures on private-sector job descriptions and how to sell themselves to an industry manager is the understanding that the private sector is a game, and they need to learn to play it. That is how I came to write this book.

In the ten years since my first book was published, I've traveled all over the world lecturing to groups of scientists who are preparing to launch their careers. The one-hour talk that I gave in the early days of my book-speaking tour has since grown into multiple short courses, seminars, and two-day workshops. I've spoken to hundreds of science PhD students and postdocs, and conversed with so many of them about their career plans and job-search challenges. I've also interviewed many industry managers to learn where the biggest gaps are in the performance of the scientists they have worked with.

In my travels, I have visited a number of universities that are doing a great job showing their early-career scientists how important and valuable private-sector careers are. Many have excellent research programs that feature successful collaborations with industry. The Optical Research Center at Southampton University and the B-PHOT program at Vrije University in Brussels stand out as great examples. But unfortunately, many universities have still not fully appreciated the importance of helping their scientists transition into private-sector careers. They continue projecting the outdated view that most scientists will become, or should become, academic research professors themselves. This view is simply not reality, and it does significant harm to continue to promote it. Many PhD students still believe they should pursue academic careers as they approach graduation, and this hurts their career planning and potential. Furthermore, the lack of exposure to mentors with private-sector experience means that few scientists learn the important principles that are critical for their success in the private sector. Through my work with TurningScience, I encourage universities and PhD advisors that they need to prepare their scientists for the careers that they will actually have, rather than continuing with the outdated view that most scientists will become professors.[1] There is still much work to do.

And in doing so, I hope we can remove the biases that promote views that either academia or industry are the only good career choices. This attitude is still far too prevalent, as illustrated by the following quote from one of the industry scientists in my first book:

> *In college, I had a professor who had a twin brother, and they had both gone to the same technical school in France. My professor continued on to an academic career, while his twin brother went into industry. The academic's perspective was that his brother may earn twice as much, but he was not discovering anything new. The one in industry looked down on his academic brother, thinking that he did work that didn't serve anyone directly.*
>
> – Antoine Daridon, PhD in Analytical Chemistry, Business Development and Marketing Manager at Metrolab Technology SA[2]

In this book, I express many opinions on both academia and industry, and these are in no way meant to disparage academia, nor suggest that the private sector is always a better career option for everyone. I've loved my career in the private sector and have known many others who found their industry careers exciting and rewarding, but I firmly believe that both segments are important to society. The private sector relies on new science coming out of academia, and academia benefits from the tools and techniques developed by the private sector.

In my work, I aim to project a better picture of both, in the hope that the participants in my workshops emerge with a better view of the need for cooperation between the two. I am also a proponent of collaboration between academia and industry.[3] Our best future is achieved with the academic sector in the private sector appreciate and respect each other, and work together. This means teaching new PhDs about both sectors, and that both sectors have tremendous value for our world. And it also means teaching PhDs how both sectors work and what working there is like, so that they can make the best-informed decisions about their own career paths.

The 'R&D Mindsets' concept that I propose in this book is intended to promote the appreciation of both the academic and industry sectors, as well as improving collaboration between the two. When scientists understand how the two worlds work and develop mindsets that enable them to be productive in both worlds, it will bring us closer to the desired outcome.

That is why I wrote this book. I hope you find it as valuable to read as I did to write.

David Giltner, PhD
Boulder, Colorado

Acknowledgments

I would like to thank Oliver Wueseke and Scot Bohnenstiehl for their very helpful input through many discussions on the topics that went into this book. I also want to thank Scot for his detailed review and critique of the first manuscript draft.

I'd like to acknowledge all the scientists and other private-sector professionals who shared their experiences with me, as well as the uncountable number of early-career scientists who have attended my seminars and workshops since I began this journey more than 10 years ago. Your stories are such an important part of this book, and your contributions will help so many scientists succeed by following your examples.

I'd like to thank my mother, Julie Giltner, for her never-ending support of my career and all that I have ever done.

Most of all I'd like to thank my partner, Eve Meceda, for her valuable support and encouragement throughout this book writing process, and in particular for being a wonderful companion through the COVID-19 pandemic, when I wrote this book.

Chapter 1
Introduction

What do you mean, 'It's a game?'

> *We're here to make a dent in the universe. Otherwise, why else even be here?*
>
> - Noah Wyle, playing Steve Jobs in the movie *Pirates of Silicon Valley*

We all set out to be scientists.

Most of us became scientists because we wanted to understand how the universe works.

Some of us wanted to use that knowledge to develop solutions that would make life better for us and our fellow humans.

This book was written for this second group of scientists.

I suspect that we all wanted to make our own dent in the universe. But as we pushed further into our early careers in science, many of us found that science research pushes slowly, that the dent it makes just wasn't big enough for us. Many of us realized that if we wanted to make a dent that we were satisfied with, we needed to hit harder and faster than simply publishing our research in scientific journals. So, we rejected the 'traditional' career path of the university professor, sometimes at the behest of our advisors, and we bravely entered the private sector. We were willing to take that risk in order to get that reward of a more satisfying dent.

Once we were in that first industry job, we learned that things work differently than in the academic research environment. We learned that success in the private sector requires different attitudes and habits than most of us picked up as graduate students. We learned that it's a game, and if we wanted to make our impact on the world, we'd better learn to play. I wrote this book to help scientists learn to play the private-sector game.

The stories I've collected through dozens of interviews with scientists working in the private sector in the last 11 years are those of scientists who wanted to shape the world. They are stories of scientists who learned to play the game and learned to win. I can't tell you 'the right way' to play the game, but I can tell you some of the important principles that these scientists and I have found to be valuable, and I can tell you many stories that bring these principles to life.

Everyone has their own approach to playing the game. We all have the same rules to follow, of course, but we all have our own playbook... our own way to win. This game-like approach can be a challenge for many of us scientists, because science is at its heart the pursuit of facts, methods, and explanations that are true for everyone around the globe. We selected a career discipline that is focused explicitly on a search for consensus. As a physicist, I am continually amazed when I reflect on how so many physical observations have been reduced to a formula that works the same way each time it is used, and as far as we can tell, everywhere in the universe.

The idea of choosing a career path where consensus is limited seems contrary to the goals that drove our selection of science in the first place. But that is what we choose when we step out into the game that is the private sector. In that game, there is far more at play than just the science involved.

> *Solving a problem in the real world has constraints far beyond just technology. That's only one of about 30 axes that matter. For mBio what really solves the problem is an adequate level of technical performance, delivered in Africa, this year, with approval from the ministry of health, and at a specified price point.*
>
> – Chris Myatt, PhD in Physics, founder and CEO of LightDeck Diagnostics (previously mBio)[1]

The Game

When I say that the private sector is a game, there are three primary elements that I consider in that analogy.

Element 1: There is more than one way to win

Most games have more than one way to win. Think of your favorite professional sport. Now think of your three favorite coaches in that sport. Do they all have the same approach to leading their teams? Do they all use the same strategy to win? They certainly all play by the same rules, and there are many principles that they agree on, but each has found a way to play the game in their own way and yet win more often than they lose. It's generally not a matter of the coach with the best record having the right formula and the others having it wrong. Next season, a different team will be the champion. Winning a game is not a formula.

And so it is with regards to working in the private sector. There are many things that may count as 'points' in your pursuit of success and many ways to consider that you have 'won.' You might be focused on salary, advancement, successfully meeting objectives, or simply a feeling of accomplishment in knowing that you made a positive difference. I can't tell you how to score each of these priorities in playing your game. You need to figure that out for yourself. I also can't tell you the 'right way' to get ahead in the game. But I can give you some strategies that have worked for others so you can write your own playbook.

> *If you could put football/soccer into a formula, it would be pretty boring because you could just run the equation and see right away which team will win. Every team would play each other once, leaving little room for learning-by-doing. Instead, pick your game and play it over and over again. That's how you get better. You're not going to get better by sitting on the sideline and trying to find the formula.*
> – Oliver Wueseke, PhD in Molecular Biology, founder and CEO of Impulse Science[2]

Element 2: Winning requires taking risks

We scientists often develop habits of over-thinking and over-analyzing. We are often inclined to keep collecting data and refining our assessments in an effort to find the 'solution,' or the way that is certain to work. But winning a game rarely comes as the result of a reliable strategy that guarantees victory. Winning requires taking a chance and not knowing whether it will work or not. Winning involves risk.

Playing games teaches you about risk. You do not win the game by stopping to carefully analyze the shot and make sure everything is just right before you take it. You win the game by creating a scoring opportunity and taking the shot. Sometimes you score, and sometimes you miss. But you learn, and you move forward with more experience and a better understanding of what takes to be a winner. You win the game by being willing to take a risk and take that shot, even if you don't know if it's the right choice or if you don't feel as prepared as you would like. You realize that now is the time, and you go for it.

And so it is in the private sector. There are rules to follow, there are players on your team, and there are players against whom you are competing. You only succeed if you are willing to make a decision when you don't know if it's the right answer or not. You only win if you take a risk.

Element 3: Knowledge alone does not make you successful

While knowledge of the rules and of winning strategies is necessary to win, it is simply not enough. At some point, you have to play the game and score. A champion sports team is well-versed in the rules, has reviewed video footage of their toughest competitors, and has a solid playbook with well-planned strategies and plays. But that is not enough to make them champions. They win because they play the game and score points.

Think of your favorite game. Are there rules that you need to follow? Certainly. Without rules, there is no game. Is it important to understand those rules in order to win? Of course. Do you become an excellent player simply because you understand the rules of play better than anyone else? Of course not. You become a better player by identifying the skills that are valuable in that game and practicing them so that you get better. And then you use those skills to score points.

> *When you play a game, you have committed to acting. If you play football/soccer, as soon as the ball is on the field, you have to accept the circumstances and act. If I sit at a table to play poker, I will have to play the cards I received. I have to assess my situation and act. I think that if you are looking for the formula, you haven't committed to playing the game yet. Scientists have to commit to playing, which goes against our training to always to try to identify the underlying formula. It's a curse of our training.*
>
> – Oliver Wueseke, PhD in Molecular Biology, founder and CEO of Impulse Science[2]

The Formula Approach

I've observed many scientists who struggle to shift into the game mentality. I struggled with this shift myself when I first entered the private sector. I've spoken to many managers who complain about their PhD scientists who 'just don't get it.' The challenge for us scientists is that we don't learn to play thc game as part of our training. Instead, we learn what I call the 'formula approach.'

From the time we begin school as small children, the academic environment teaches all of us a formulaic approach. We learn that if we answer all of the test questions correctly, we will get an 'A' grade. Get an 'A' on every test, and you'll get an 'A' in the course. By the time we go to college, we find that if we pass all of the classes in a particular checklist, the university will give us a degree in that particular major. Those of us with PhDs spend more time in the academic environment than anyone, long past when most other people were already learning how to play the 'real world' game. By the time we graduate with our PhDs, we are well-versed in this habit of thinking.

Playbook story: Choosing the low-risk game

I had a unique opportunity to reflect on games and the risk involved several years ago when I attended a basketball game with a friend to watch her sons play. I tried out for basketball in elementary school because that's what most of the guys in my school did, but I was quickly cut from the team due to my lack of skill at the game. My athletic pursuits since then have all been solo activities where I compete primarily against myself, such as trail running and cycling, so the opportunity to watch these young boys playing a team sport was a new experience for me. It occurred to me that they were willing to run down the court and throw the ball up for a long shot without stopping to think about it, despite the significant risk of failure. They were willing to take a shot in front of an entire room full of spectators, knowing that they might miss and half the audience would sigh and groan at their failure. This really impressed me, and I realized that I had not developed that strength at such a young age.

This realization caused me to reflect on the games that I enjoyed when I was young. My favorite game was 'Mastermind.' It was a game for two people, where one player creates a series of four colored pins selected from six colors, and the objective is for the second player to determine the sequence through a series of guesses. Player two chooses a four-pin 'guess,' and player one indicates if any pins are the correct color only or the correct color and location. Through a series of such guesses, player two can figure out the sequence assembled by player one.

I enjoyed this game very much as a child, and still do. Looking back, I realize that I enjoyed it primarily because it is possible to figure out a process, or you might say a formula, for guessing the sequence in no more than five guesses.[3] In this way, one can use brain power to reduce the uncertainty in the game. I enjoyed being able to remove the uncertainty and reduce the risk of losing.

I also realize now that the uncertainty in basketball is one of the reasons I did not enjoy the game in elementary school. Because I did not have the skills needed to be good at it, the uncertainty in my performance was far greater than that of the average player, and I was not willing to take that risk. Rather than improve my skills, I decided basketball was not for me. But that meant I did not develop that appetite for risk that I believe the guys who continued playing did.

In college, I chose a path where I could leverage brain power to find the right answers to most challenges and largely avoid the need to 'take a shot' that I might miss in front of a crowd of onlookers. Tests can be studied for and presentations can be rehearsed in front of friendly audiences to greatly increase the chances of success. This risk-taking stuff was something I'd largely avoided—that is, until I started my career in the private sector.

I suspect that many of us became accustomed to using our intelligence to deduce the formula for success and remove the uncertainty. If you want to be successful in the private sector, however, I encourage you to embrace the uncertainty and learn to take risks.

Our training as scientists only reinforces this formula approach. The goal of science research is to develop theories that we can use to predict outcomes. We are searching for determinism. And many of us have become comfortable working in the space and working toward the expectation of predictability.

We develop the expectation that with enough data and analysis, we can find the 'right answer.'

The problem is that life outside the science lab doesn't work this way and therefore neither does building a career in the private sector. This is where scientists looking for careers in industry first meet head-to-head with this different perspective and begin to struggle. Many newly minted PhDs ask me 'If I want to get a job doing __________, what other courses should I take or what certifications should I get?' This is formulaic thinking, and it is not the way to build a winning career in the private sector. If you have a PhD, you've already had more formal training and education that most people on the planet. It's time for you to get in and play the game!

If you find yourself taking this formulaic approach to your career, this book will help you learn to think of it as a game. Identify your strengths, develop systems to improve your strengths, and learn to tell stories that position you as a player who can help that company you have your eyes set on.

The Value of Stories

I use lots of stories in my work, and so this book includes many stories. Several of them come from my own career, but many more come from the dozens of industry scientists, industry managers, and company executives who I've interviewed. Others come from the startups that I've coached through my entrepreneur mentoring activities, and still more come from the hundreds of PhD students and postdocs I've worked with in the lectures and workshops that I give to early-career scientists around the globe. My hope is that the stories support the principles that I describe in the book and will give you a feel for the game.

Why stories? Stories draw you in and make you part of the experience. Rather than simply listing bullet points that I hope you figure out how to implement in your own career, a story helps you visualize how a particular idea worked for someone else. A story helps you internalize the idea and decide how to use the information for your own career.

Also, stories are much easier to remember than paragraphs describing ideas on how to be more productive. Humans have used stories since the beginning of recorded history as a way of passing information down from generation to generation. This was done largely because it was a good way to make a principle memorable.

My favorite books on business or productivity are full of stories, because I find them so much more relatable than theory or a list of bullet points. A story is so much easier to remember while I go about my life looking for the right way to work the underlying principle into my own playbook. I hope you find the stories I tell in this book to be insightful and help you develop your own winning playbook!

Playbook story: Maurice Hilleman and winning the game

One of the most significant contributions to the health and lifespan we currently enjoy was made by a scientist who ignored the pleas of his professors that he stay in academia and instead launched his career in the pharmaceutical industry.

Maurice Hilleman was born in Helena, Montana in 1919 and received his PhD in microbiology from the University of Chicago in 1944. He completed his postgraduate studies with a singular focus: creating vaccines to protect humans against some of the most debilitating diseases of the early 20th century. He was focused on making a difference in the world, and he realized that he could not make enough of an impact working in an academic research environment. If he really wanted to provide a significant benefit to humanity, he needed to work in industry. Paul Offit tells Hilleman's story in the book *Vaccinated: One Man's Quest to Defeat the World's Deadliest Diseases*:

> "In 1944 Maurice Hilleman came to a crossroads. He had just finished his graduate studies in Chicago. Now he was expected to take his place among the academic elite as a teacher and researcher. But Hilleman wanted to work for the pharmaceutical company E. R. Squibb in New Brunswick, New Jersey. His mentors made it clear to him that working for a pharmaceutical company wasn't an option. 'I left Chicago under significant duress,' recalled Hilleman. 'Because Chicago at that time was such an intellectual center of biology, no one went into industry.' But Hilleman had tired of academia. 'What am I supposed to do? I was told that you could teach or you could do research. I said I wanted to go into industry because I'd learned enough about academia. I wanted to learn something about industrial management. I came off a farm. We had to do marketing. We had to do sales. I wanted to do something. I wanted to make things!' Hilleman's decision irked his professors."[4]

In 1957, Hilleman moved to Merck & Co, where he spent the next 47 years as the head of the new virus and cell biology research department. He is credited with developing more than 40 different vaccines during his career, including measles, mumps, hepatitis A, hepatitis B, meningitis, pneumonia, *Haemophilus influenzae* bacteria, and rubella.

His vaccines have been credited with saving millions of lives and eradicating multiple childhood diseases. The measles vaccine alone is estimated to have prevented approximately 23 million deaths worldwide between 2000 and 2018.[5] When Hilleman died in 2005, scientists in the field credited him with likely saving more people than any other scientist in the 20th century.[6]

Risk

In 1957, Hilleman took bold action after reading early reports of an influenza outbreak in Hong Kong. Realizing what was likely to follow, he took a risk and proactively invested in developing a vaccine before it spread:

> "In April 1957, a mysterious illness was making its way through Hong Kong. Medical workers encountered throngs of children with 'glassy-eyed stares,' and more than 10 percent of the city's population was infected with influenza. The scientific community stayed quiet, but American virologist Maurice Hilleman recognized the threat: A pandemic was brewing."[6]

Hilleman realized that this disease was likely a new strain of influenza that could be capable of spreading around the world. Rather than follow the official chain of command through his employer, he contacted a U.S. Army lab in Japan and convinced them to collect a sample from a Navy serviceman who had become infected. A few weeks later, the sample arrived at Hilleman's lab, and he and his team began 14-hour days to isolate the virus strain.

Once the virus was isolated, Hilleman sent samples to six U.S.-based companies that already produced influenza vaccines. He knew that if there was any hope of saving lives, he had to persuade these companies to create and distribute the vaccine in only a few months:

> "To (get the vaccine quickly), Hilleman ignored federal drug regulators: 'I knew how the system worked,' Hilleman said. 'So I bypassed the Division of Biologics Standards, called the manufacturers myself, and moved the process quickly.'"[7]

His boldness and risk-taking paid off big:

> "By the time the virus arrived in the U.S. in fall 1957, he was ready with a vaccine. His work prevented millions from contracting the deadly virus—and that's a small fraction of the people Hilleman would save over the course of his career."[6]

While thousands of Americans still died from influenza in the year that followed, the number of deaths was greatly reduced due to Hilleman's willingness to take a risk. This accomplishment was tremendous, as it was the first time ever that a pandemic was averted with a vaccine:

> *Whenever you do something like that, you have to be (a) pretty confident about what you're doing, and (b) pretty ballsy about it. And I think Maurice Hillman had both of those elements in his personality. He could've turned out to be wrong. He wasn't.*
> – David Oshinsky, Pulitzer-Prize-winning author of Polio: An American Story[8]

Winning

> *For a scientist it's the winning that counts. It's like climbing mountains. You get up to the top of this one, and you've got a couple more that you're trying to climb at the same time, and it's an interesting kind of thing. Yes, there is personal satisfaction in doing something. This all goes back to that ethic of doing something useful and being useful to the world.*
>
> – Maurice Hilleman

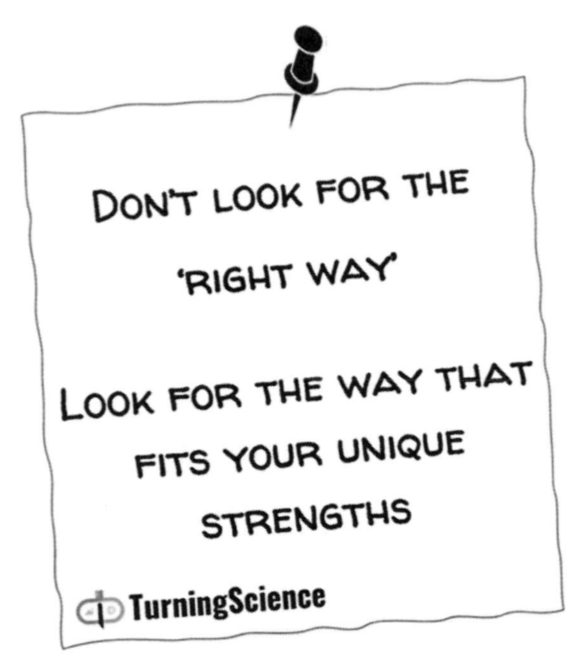
DON'T LOOK FOR THE
'RIGHT WAY'
LOOK FOR THE WAY THAT
FITS YOUR UNIQUE
STRENGTHS
TurningScience

Chapter 2
Rules of the Game

How is industry different than academia?

Each of my private-sector career workshops starts with an interactive session that asks the participants to answer four questions. Their answers to these questions help me understand their backgrounds and direct the workshop more effectively. One of these questions is, 'How is working in industry different than working in academia?' From their responses, I've learned that many science PhD students are aware of a few of the major differences between the two environments. Most of them are aware that industry is about generating revenue, and many of them have heard that the pace of progress is typically much faster in the private sector than in academia. But beyond this, most PhD students don't know much about what industry is really like.

Of course, neither did I when I was a graduate student. When I decided during the last year of my PhD work that I wanted to build my career in industry, I realized quickly that I knew almost nothing about the environment where I decided to build my career. Because nearly everyone in the physics department had spent their entire careers in academia, there was no one to help me with my career planning. I remember the conversation I had with my advisor about my plans to work in industry, when she said, 'If you want to do a postdoc, I can help you. I have lots of connections and can help you get a great position. But if you want to go into industry, you are on your own. I don't know anything about it, and I don't know anyone who works in industry.' My advisor wanted to be helpful, but like so many successful academic researchers, she had absolutely no experience with the career path I had chosen.

Finding a job with no guidance was a big enough challenge, but once I started that job, the next few years were a very steep learning curve for me. I began my career at SDL, Inc., a company in San Jose, California that was developing semiconductor laser technology for a variety of applications. I was fortunate that they employed many PhDs, due to the cutting-edge nature of the semiconductor laser technology and the significant amount of research that was required to bring it to the level of a commercial product. Most of the

THE RULES

PRINCIPLE:	ACADEMIA	INDUSTRY
WHAT IS CREATED:	KNOWLEDGE	PROFIT
WHAT IS PURSUED:	UNDERSTANDING	RESULTS
WHAT IS REWARDED:	CERTAINTY	SPEED
HOW ONE SUCCEEDS:	INDEPENDENCE	INTERACTION
HOW PROGRESS IS MADE:	PROOF	PERSUASION

Playbook Sketch 1 The rules that govern academic research and industry.

upper-level managers at SDL were PhDs who had moved into leadership roles, so there was a strong culture of helping beginners to learn the game. Despite this, it still took several years for me to fully realize how my academic training had instilled habits in me that no longer worked in this new environment. Many scientists who are new to the private sector do not have the advantage of great mentors, and so their struggle is often even greater.

If you're going to play this game effectively, you need to understand the rules. Therefore, the best place to start a discussion of the private-sector game is to outline the rules. I like to describe the industry working environment in contrast with the academic research environment. The differences between the two environments have to do with the primary goals in each and how these goals are met. I will refer to them as 'principles,' but one might think of them as 'rules of the game.' While there are exceptions, these are the principles that govern most of what happens in each of the two environments. The first principle is the most fundamental, and we find that the other four actually follow from the first one (see Playbook Sketch 1).

Difference #1: What Is Created

Academia: Knowledge

Science is the pursuit of a better understanding of the universe, and this pursuit results in the creation of new knowledge. The measure of success in an academic career is typically considered to be getting research grants and publishing in scientific journals.

Industry: Profit

Above all, companies exist to make money. Behind every company are one or more investors looking for a return on the capital they have invested. This means that no matter what kind of wonderful world-transforming products a company may produce, the primary focus of the executive team is to earn a profit.

After some of my seminars where I describe this primary difference between academia and industry, I've had professors approach me and say, "But companies aren't the only ones who have to worry about money. Professors have to worry about money, also!" This is true, but it misses the underlying point: industry isn't just about money, it's about profit.

When money is spent in the private sector, it is usually seen as an investment, not simply an expense. Researchers apply for grant money that they can spend to do the work that will result in new knowledge. Successful business executives think about how to spend the money they have in such a way that brings a positive return, i.e., so that their investment earns more than is spent.

Academia spends money
Industry ***invests*** *money*

Difference #2: What Is Pursued

Academia: Understanding

The primary objective of academic research is a full understanding of the phenomena being studied. When a research team submits a paper for publication, they take care to demonstrate the completeness of their work, as any apparent lack of a full understanding is likely cause for rejection.

When a graduate student defends a thesis, the committee reviewing the work will ask detailed questions designed to verify that the student has developed a sufficient understanding of the project. A wise PhD candidate prepares for every anticipated question in order to ensure passing the defense and receiving a degree.

This pursuit of a thorough and complete understanding is fundamental to academic research. As such, research projects often progress slowly over many years. In some cases, entire careers are built on the pursuit of a thorough understanding of a particular topic.

Industry: Results

Results are generally things that ultimately produce either revenue or reduced costs, both of which create profit. Examples include a product or service that can be invoiced, an improvement in final test yield, a reduction in the cost of a product, or simply a solution to a problem that is unnecessarily costing the company money.

Because the results that are pursued are intended to produce profit, this principle is derived from the first one listed here. And because achieving results quickly and efficiently ultimately results in more profit, the pursuit of results often comes at the expense of a complete understanding.

Any effort beyond what it takes to achieve the desired result is wasted time and money, so desired results are defined as clearly as possible. Specifications are the most common way to do this, as they define what is 'good enough' from a technical standpoint. An effective team works to achieve the specifications and then quickly moves on to the next project. A scientist who spends additional time to gain an understanding that is not required to meet the specifications is reducing the efficiency of the team and ultimately hurting the profitability of the company.

When I was a graduate student, I felt it was important to work out every derivation for myself so that I would have a complete understanding of how that result was obtained. I was not happy to simply take a result derived by someone else and use it to solve problems. I wanted to intuitively understand where the result came from. This is one of the great benefits of being a student. Postgraduate education provides the time to develop that complete understanding. It hones your intuition and develops habits of thinking that will benefit you for the rest of your career. But while the benefits of this habit may last your entire career, the habit itself likely should not.

When you transition into industry, you will no longer have this luxury to spend hours or weeks developing a compete understating. Working to re-derive a solution that your colleague derived might be labeled 're-inventing the wheel' and frowned on by your manager.

Difference #3: What Is Rewarded

Academia: Certainty

Academic researchers place great importance on the accuracy of their work, so they review their research findings thoroughly before they submit them to be published to ensure that there are no mistakes and that their conclusions are correct. The peer-review process that is part of the publishing cycle is yet another check on accuracy, as is the effort by other researchers to reproduce key research findings, a fundamental part of the scientific method. Publishing inaccurate conclusions can result in a significant loss of credibility for any scientist.

Industry: Speed

The private sector is a fast-moving game. Any company selling products and services that people actually want will have competition, so moving quickly is paramount. This doesn't mean that certainty can be ignored, of course, as results in industry do ultimately need to be correct. Qualification test results need to meet the accuracy that is claimed. Specifications and reliability targets must be met by the time a product is shipped to a customer. However, many projects move forward based on the probability that they will work, not based on conclusive results. Certainty is achieved over time as the project progresses.

Specifications are used to define the level of certainty or accuracy that is required, and they are developed with the perspective that usable results next month are much better than perfect results two years from now. A scientist who spends additional time working to achieve a level of certainty beyond what is required by the specification is wasting time and money, and ultimately hurting the profitability of the company. If a customer desires a higher level of certainty or accuracy, they will need to pay for it so that the additional investigation time is justified.

In my workshops, I often say that "Everything in industry is done as a project with a plan, a schedule, and a deadline." The deadline for a project is important because other projects are likely waiting on the results, and delays in one project will cause delays in others. Therefore, most projects include intermediate milestones to allow the team to determine if they are on track long before the deadline is reached. This allows adjustments to be made so that the program stays on schedule.

Detailed project plans of this sort are far less common in an academic setting. In my time in academic research, the only deadlines I ever saw were for conference paper submissions. In industry, it's all about moving fast based on what you expect is likely to be the result and then figuring it out as you go.

> *I was shocked by the time pressure in industry, where a very long project might take a year and often much less. Projects take much longer in an academic research environment. A PhD takes many years, and you are expected to take your time, look at every single detail of your problem, and know it inside and out. In industry, it's the bottom line that counts, and it's okay if you don't understand every little detail of your product as long as it works properly. It is only when it doesn't work properly that you invest the time to go back and understand it better.*
>
> – Ashok Balakrishnan, PhD in Physics, CTO and co-CEO of Enablence Technologies[1]

Difference #4: How One Succeeds

Academia: Independence

The quintessential image of a successful professor is an independent researcher who has developed an area of expertise where they are considered the authority on the subject. Success in academia is typically achieved by focusing one's career on a particular research area and developing an expertise in this field over time. For some scientists, this is a major consideration in their career choice:

> *(What) bothered me (about an academic career) was the focus on being an independent silo. It seemed that in academics, you are lauded if you focus completely on one thing. You become the person who does 'X,' and you're the best in the world at doing 'X.' That's how you get research grants and tenure and get to give talks around the world, but it seemed to be very limiting. I also felt it requires you to step on the graduate students below you, because you aren't actually the one doing the work. The graduate students really don't even want their advisors to touch anything in the lab, so the professor doesn't actually do any of the lab work, but they have to take credit for all the work the graduate students do. That didn't sit well with me.*
>
> – Kate Bechtel, PhD in Physical and Analytical Chemistry, Biophotonics Fellow at Triple Ring Technologies[2]

The importance of being an independent expert is driven largely by the incentives, which are based mostly around securing research grants and university appointments. The Nobel Prize, considered by many to be the pinnacle of an academic research career, is given to those who, "during the preceding year, have conferred the greatest benefit to humankind."[3] Despite the language in Alfred Nobel's will that outlined his intentions, the scientific prizes have typically been given to reward contributions over an entire career rather than a single year. And while not explicitly intended for a researcher who has spent their career developing an expertise in a particular area, this description generally applies, as achieving the intended level of impact usually requires an extended focus on a particular area of expertise.

Recognizing the 'benefit to humankind' of a researcher's work often requires many years after the rewarded discoveries are made. This is in part because awards committees have historically rewarded discoveries over inventions by a factor of more than 2 to 1.[4] As an example, 77% of Nobel Prizes in physics have been given to discoveries, compared with only 23% to inventions.[5] This is because discoveries tend to be more highly regarded by the scientific community than inventions. Whereas an invention addresses a specific need, such that the benefit might be recognized quickly, a discovery typically does not, and realizing the benefits to humanity can take time.

The inclination for a scientist to become an expert in a particular field goes beyond the possible career incentives. The very nature of academic training encourages us to be the independent expert and have all of the answers. From the time we enter elementary school as young children to the time we complete the coursework for our postgraduate studies, we learn that we need to have all of the answers if we are to get passing grades. Completing a PhD only strengthens this inclination, as we are given a project that we are expected to complete independently, after which we will stand in front of a thesis review committee and defend our work without anyone else to assist us.

The result is that, through a lifetime of training, we scientists develop a habit of striving to become that independent expert and know all the answers if we are to be successful. This makes sense, given the goals of a PhD research project and the independence that being awarded such a degree signifies. The prestige and accomplishment that it confers is important and valuable, but it is not well-aligned with the way success is achieved in an industry setting.

> *I've found that, in some ways, graduate school prepared me for the opposite of the real world. It taught me to work alone.*
> – Scott Fulbright, PhD in Cell and Molecular Biology, CEO and co-founder of Living Ink[6]

Industry: Interaction

The academic model of the independent expert is not consistent with the private sector's focus on fast results. When faced with a problem, it is far faster to find someone else who already knows how to solve it than it is to learn how to solve it yourself. Therefore, companies build teams of people with diverse skill sets rather than look for someone who can singlehandedly accomplish the task. In industry, it's the team that gets the credit, not an individual who accomplished everything alone.

Successful leaders in industry don't try to be the experts.
They become skilled at finding people who can help them.

When I managed a product-development team earlier in my career, my attitude was that I needed the people on my team to find the answers, not necessarily to know the answers themselves. If they recommended that we hire a consultant to help solve the problem, because they were not familiar with the specific issue, that was fine. I told them I'd much rather pay someone to get the answer in a few days than wait three months while they try to figure it out themselves.

In addition to enabling faster progress, the team approach allows more flexibility than the expert approach. Companies will frequently pivot into a new market or technology area where new expertise is required. The scientist who has invested time in learning everything there is to know about a particular area may quickly become less valuable than intended. The scientist who immediately thinks, "who can help me figure this out?" will be much more valuable.

When I discuss this particular difference between academic research and the private sector during a seminar, I'm often reminded by university researchers that current trends in academic research involve much more collaboration and teamwork than they did a couple of decades ago. This is true, and it is a wonderful trend. However, the primary distinction remains. An aspiring professor needs to be well-versed on every aspect of building an independent research program, including the technical details of their research, securing funding, managing a team of students and postdocs, and navigating university bureaucracy. This degree of independence required to become a successful PI is a rare requirement for building a career in the private sector. It is worth noting that although a university professor will have specific research interests, the overall skill set will be remarkably similar to that of the professor's peers. This is quite different than the situation when one works at a company. In industry, one is likely to have very different skills and strengths than most of one's peers.

> *Working in teams to solve scientific problems is an essential skill many PhDs are lacking. This lack of experience causes significant transition hurdles when moving into industry, where these skills are part of everyone's tool kit. Without them, your career progression in industry can be slowed down. You'll have to learn these skills on the job, which means a company will have to invest time and money into your education.*
>
> – Oliver Wueseke, PhD in Molecular Biology, founder and CEO of Impulse Science[7]

Difference #5: How Progress Is Made

Academia: Proof

Fundamentally, science is about confirming hypotheses with sufficient data to be highly confident in the results. Technically, the scientific method is a process of trying to disprove a hypothesis rather than prove it, but this typically involves collecting a large amount of data that supports the hypothesis. Certainty is the objective. A researcher will generally continue collecting supporting evidence until they feel certain that the hypothesis is validated before they are willing to publish the results. This is the method by which the 'certainty' described in principle #3 is achieved. In the academic research lab, certainty is achieved with lots of data and analysis, which is often time consuming. Speed is traded for certainty, and the result is that science research tends to progress slowly.

Industry: Persuasion

Industry is all about achieving results quickly. Only when a decision is made can a team make progress on one of many possible paths to achieve the desired results. The need for speed means decisions often need to be made without the luxury of certainty in the outcome. When proof is not possible, one relies on persuasion to move forward.

Data is important in the private sector, but many decisions are made without enough data to prove what the right decision is. There are two primary reasons for this. First, there is usually not enough time to collect a sufficient amount of data to be certain. If the team waits until the scientists collect enough data, the competition will have gained the lead and/or the customer will have moved on to another solution. Second, in many cases, no amount of data will show what the 'right answer' is.

Unlike the science lab, where researchers generally try to isolate a particular phenomenon so that it can be explained and predicted reliably, decisions in the private sector involve many additional considerations. Most of these additional considerations cannot be reliably predicted. These may include the future plans and decisions of their customers or of vendors who supply critical materials for their products. It may be that the value proposition of a new product depends on information that is kept private by the members of their target market. It may include unpredictable macroeconomic conditions or perhaps the unforeseeable impact of a global pandemic. And it certainly includes the fact that every company is a collection of human beings, each with their own private life that is impossible to predict. Rather than having too little time to determine the 'right answer,' the issue is that it is simply not possible to predict the outcome of most decisions in advance. It's best to consider that there is no 'right answer' to begin with.

Successful leaders in the private sector learn to make decisions quickly with limited data, and therefore limited certainty. They recognize the value of evidence, and they strive for 'data-driven' decisions when possible, but they realize that there is a limit to how much data can realistically be collected before they need to make a decision. They learn to move forward based on the probability of the outcome, rather than certainty. They work to get the odds in their favor and then move quickly.

And in an environment where you don't have enough data to prove your decision is the 'right' decision, you must persuade others to follow you. This is very different than academic research, where a scientist submits a paper for publication and expects that the referees will clearly see the merit of the paper and its conclusions based on the quality of the data and the analysis. Influence in academia is based on the quality of one's work, not the quality of one's argument. But in an environment where there is rarely enough data to do the talking, the scientist must speak for it. Persuasion and influence are critical to making progress.

Scientists who can make decisions with limited data and then persuade others to follow them are the ones who make the most progress. These are the scientists who move into positions of influence.

If you are going to a meeting and you want things to go to a certain direction, you never introduce your point for the first time in the meeting. You get far better results if you talk to the key people before the meeting and bring them on board with your vision. You need to sell your vision in advance so that by the time you get to the meeting, the decision is already made.

– Yasaman Soudagar, PhD in Physics, co-founder and CEO of Neurescence[8]

Scientists and the Game

These are the five principles that most clearly differentiate academic research from work in the private sector, and they outline the 'rules' of the private sector game. The question that follows is 'Are scientists good at playing this game?' My own experience has led me to believe that most are not naturally good at it when they first transition into the private sector. I've confirmed this suspicion through numerous interviews and informal conversations with experienced industry managers. Our PhD training teaches us the 'formula approach' outlined in the first chapter of this book, and few of us enter the private sector aware of the game environment we are entering. This leads to the PhD stereotypes described in the next chapter that describe how so many non-scientists in industry view the PhD scientist.

Academic researchers reading this chapter may ask, 'Isn't academia also a game?' I would agree that academia is a type of game as well. I am pretty sure that most experienced researchers would consider the pursuit of funding to be a very challenging and competitive game.[9] However, the majority of scientists transition into the private sector early enough in their careers such that they never have the responsibility of securing funding, so they generally stay in the 'formula approach' mode until they get their first industry job, and they rarely experience much of the game aspects of academic research.

Author's note: Why is academic research not the 'real world?'

I often use the term 'real world' for industry/private-sector work. Every now and then I meet someone who seems offended by the use of that term, so I want to explain myself. After all, isn't science research always performed in the real world? Well, of course it is. The whole goal of science is to figure out how the real world works. And of course, a science lab exists within the real world.

But the whole purpose of science is to isolate some aspect of the universe and remove as much uncertainty as possible so that we can distill a certain behavior into something that is predictable and can be described with a theory or model. This is perhaps most clear in physics, where theories are expressed by mathematical formulas. Think of one of the earliest problems you likely solved in physics, involving a block sliding down an incline plane. This problem is an abstraction of the real world and begins with implications such as neglecting friction and variations in density and surface quality. A more humorous example is the 'spherical cow' metaphor that physicists often use to poke fun at the highly simplified mathematical models theorists may use for complex, real-life phenomena.

The point is that science is done in a controlled environment, where we try to isolate the specific phenomenon we are studying from the uncertainties of the environment where we actually experience them. A technically based company is focused on using that scientific knowledge to create a product or service that helps humanity in some new way, and this means bringing many uncertainties into the equation. The result is no longer deterministic and is rarely fully predictable.

Science may choose to ignore uncertainties such as friction and turbulence, but companies are advised not to ignore uncertainties such as the unpredictability of their customers' businesses or the fact that each of the human beings in the company has their own unpredictable personal life. This is the real world, and this is the environment in which decisions need to be made when running a company.

So when I use the term 'real world' to describe the private sector, it is not meant to disparage academia or the nature of science research. It is simply an expression of the fact that science attempts to look for predictability, but the private sector operates in an environment that is inherently unpredictable. This is the playing field for the private-sector game. It is unpredictable and unforgiving, but it is an exciting and rewarding game to play for those with a sense of excitement and appetite for risk.

KNOWING THE RULES IS GREAT...
BUT NO ONE BECAME A CHAMPION
PLAYER BY KNOWING THE RULES BETTER
THAN ANYONE ELSE
THAT'S HOW YOU BECOME A REFEREE
TurningScience

Chapter 3
The PhD Stereotypes

How are scientists viewed by the private sector?

An Awakening

I didn't always plan a career in industry. In fact, for most of my academic career, I assumed I would become a professor. This was primarily because it was the traditional route for a scientist, and it just seemed like the natural career to follow when you have a degree in science.

But after five years of watching my advisor and the other professors in the physics department, I decided that career was not for me. I didn't like the idea of working in the same building, in the same office, with the same colleagues for the next 40 or so years of my career. I wanted something with more variety. Also, as much as I enjoyed working on fundamental physics research, I was beginning to feel a lack of satisfaction with simply producing publications that only a handful of people around the world would ever read. I wanted to make something tangible that people might use in a few years. I wanted to see the results of my work. So, I decided to head into the private sector.

I found my first job at SDL in San Jose, California, where my first task was to assist another young PhD physicist developing a laser system using semiconductor laser technology for a Small Business Innovation Research (SBIR) contract awarded by the U.S. government. Semiconductor lasers were a relatively new technology at the time, so the work involved a lot of experimentation to determine their capabilities. The intention was that the technology we were developing would ultimately be turned into scientific instruments sold to science labs as commercial products. However, in those first few months the work felt much like my PhD lab work, as I was just trying to get an experiment to work properly.

However, it didn't take long for me to realize that this private-sector environment I had entered was different than academic research in some very important ways. We had a paying customer, so this project was no longer a matter of working to get the results I thought were best. I needed to achieve the results that the customer was paying for.

And while it was wonderful that I was being paid a lot more money than when I was a graduate student, extra money also meant my time was much more valuable, and I need to think seriously about the efficiency of my work. Gone were the days of spending weeks in the lab trying to answer some novel and interesting question. I needed to work fast to demonstrate results for the customer.

But those weren't the only differences I noticed and certainly not the most challenging to accept. I soon came to realize that the people who get the most attention in a company are the ones who bring in revenue, not the 'smart and highly educated' ones.

As a newly minted PhD, I thought of myself as intelligent, independent, and highly skilled. Now I was working at a high-tech company in Silicon Valley, helping them develop products that would change the world! But I soon realized that I was just a guy who was trying to design something that would someday, hopefully, bring in revenue for the company, and in the meantime cost the company a lot of money. That's not quite the prestige that I had imagined.

My first job was a welcoming environment, because the company hired a lot of PhDs due to the cutting-edge nature of the technology we were developing. But as I moved out of the lab and into a variety of leadership and customer-facing roles, I began to see more clearly that not everyone saw my PhD the way I had. I'd wanted to see myself as one of the more important people in the company, but people with more business experience did not always see it that way.

I slowly realized there is a PhD stereotype in the private sector and that some of my attitudes and behaviors fit that stereotype. That was an uncomfortable realization and a real 'wake-up call' for me. If I wanted to be well-respected by senior management, I needed to quickly develop some new thinking and working habits.

I now tell my private-sector career workshop participants that the CEO likes the sales team better than they like the scientists, and I'm only half joking when I say it. CEOs have to worry about profit, and it's the sales team that actually brings in the orders so everyone can get paid and the company can remain profitable. As a PhD scientist, we hope that our work will turn into profitable technology at some point in the future, but there is no guarantee of that.

This realization was the start of a larger understanding that the private sector is a game, and my PhD training did not teach me to play it.

The Challenges for PhDs in Industry

In designing my industry-career workshops, I've interviewed a number of industry managers to understand what some of the successes and challenges are for PhDs moving into industry. I was certainly aware of some of the stereotypes, but I wanted to get it straight from people I trusted. Often, the managers were quicker to mention the challenges than the successes!

I've also investigated the struggles PhD job seekers face, curious how the hiring managers' views might impact their hiring decisions. I've found that the job seekers do in fact encounter hiring managers who view PhDs based on these stereotypes.

Let's start with what the job seekers observe:

The job seeker's perspective

Many PhD scientists looking for careers in the private sector observe resistance to hiring PhDs from the companies they pursue. The PhD Plus 10 Study conducted by the American Institute of Physics tracked the careers of physics PhDs 10 years into their careers to learn about the success factors and barriers to PhD scientists building private-sector careers.[1] Many of the interviewees reported encountering negative attitudes toward PhDs when applying and interviewing. Below are some of the quotes reported:

> *The general perceptions of academic science that accompany a doctorate in physics are a huge barrier. I was once told by a manager in the aerospace industry that I had 'no useful skills for the aerospace industry and I should go find a place with white lab coats.'*
>
> *A physics degree is seen as an overqualified 'brain in a jar' by the business world.*
>
> *Many feel I'm 'too smart' to even hire me for a job in the first place.*
>
> *Having a physics PhD was a hindrance in my early career. I asked an acquaintance, once, why I wasn't getting any responses from my job submissions. I was told, point blank, that many companies don't want to bring on PhDs as they can be 'too disruptive.' It is perceived that they will be difficult, too smart, or too bored.*

I've heard about similar experiences from the scientists I interviewed:

> *I found that in job interviews, people would look at my resume and think that because I had a PhD, I would be very research oriented or too much of a nerd. You do have to convince people that's not who you are. You have to convince them that you can work with other people and you are interested in actually solving the problems that your employer faces, not just in solving problems for the sake of solving problems. You have to*

convince them that you are practical and are not going to take forever on a project.

– Ashok Balakrishnan, PhD in Physics, CTO and co-CEO of Enablence Technologies [2]

The perceptions that the candidates received from their job searches are almost certainly not the complete story. It is unlikely that a candidate would truly be considered 'too smart' for a job, as intelligence is considered an advantage in all aspects of business. The claims of 'no useful skills' and 'too disruptive' may hold more helpful clues as to the real reasons for their struggles. To get a more complete perspective, it's helpful to speak directly with managers who have worked with PhDs on their staff.

Author's note: What a PhD meant to me

There are many different reasons why each of us pursued an advanced degree in science. Some of us grew up with parents who were scientists, and we decided to follow in their footsteps. Some of us had a wonderful science teacher early in our academic career and that experience inspired us to want to teach science to others. Some of us became fascinated with how the world around us worked at an early age, and the traditional science education path naturally led us to a PhD.

I was in this last category. I declared a physics major early in my undergraduate career because I enjoyed the physical sciences, and the coursework came easily to me. While earning my undergraduate degree, I realized that one doesn't truly become a scientist without a PhD.

But when I'm honest with myself, I admit that there was another very important reason I got a PhD: I wanted to prove that I was smart. I wanted to prove it to myself, and I wanted to prove it to others. Once I'd completed a PhD, I'd forever have those three letters behind my name that proved to everyone for all eternity that I was smart.

As I advanced my career in industry, I learned that this attitude was actually a liability. This realization caused me to re-evaluate some of my career choices briefly, but I decided that both the PhD and the career in industry were great choices for me. I just needed to shift my attitude and learn to play the game. Academia had trained me in some very valuable technical skills, but it had taught me very little about how things get done in the real world. To many people in the private sector, I was just a stereotypical PhD.

The hiring manager's perspective

What could be the issues that industry managers see that are so significant that they offset the strengths that a PhD scientist brings? Here are two representative quotes from industry managers I've interviewed:

We have hired quite a few PhD scientists over the years, but we generally don't have as much success with them as we do with masters- and bachelors-trained people.
– Tom Baur, MS in Astrophysics, founder, CTO, and chair of the board of Meadowlark Optics[3]

We have a number of PhDs on the team, but they just don't understand what we need from them. We need results, not endless analysis. We don't hire many PhDs anymore.
– R&D manager in the medical industry

The sentiments expressed by these managers are very unfortunate, given the valuable strengths possessed by PhD scientists in all disciplines. However, in more than 20 years working in industry myself, I've observed many of the challenges that PhD scientists face when transitioning into private-sector careers. Of course, these comments also do not apply to all PhD scientists, as there are many who are successful in the private sector.[4,5] Many even go on to become executives in successful technical companies. But the problems are common enough that they leave an impression on managers and result in a hesitancy to hire PhDs, who frequently see the gaps themselves once they get their first private-sector job:

I felt totally unprepared! And rightfully so, because I was unprepared! Nothing in my previous education gave me much exposure to the product-focused mindset that is the norm in industry. I had to engage in an accelerated crash course training on the job, which meant talking to people, trying, failing, and waking up in the middle of the night because I thought I had screwed up.
– Oliver Wueseke, PhD in Molecular Biology, founder and CEO of Impulse Science[6]

What is the difference between a successful and an unsuccessful PhD scientist in industry? It has more to do with behaviors than with intelligence or skills.

Author's note: Am I overqualified?

A concern that I frequently hear from PhD scientists who are looking for their first job in the private sector is, 'I've heard from a few companies that I'm overqualified. Is that true?' When a PhD scientist hears this excuse from employers, it's usually for one of three reasons:

1. They were competing for a job that did not require the level of independence and training that they possess (in this case, they are overqualified).
2. They are positioning themselves as experts in a niche research area and are missing the real strengths they bring.
3. They come across as the stereotypical PhD scientist, and 'overqualified' is the euphemism that the employer used when turning them down.

The strengths of a PhD scientist

One of my interests is the strengths that make PhD scientists most valuable in the private sector. I regularly survey the scientists in my workshops on the most valuable strengths of a PhD scientist. Four of the top responses I've collected are the following:

1. The ability to apply critical thinking and identify and solve problems that others don't see.
2. The ability to work independently and determine what needs to be done without a lot of supervision or support.
3. Our intelligence and the ability to learn new things on our own without explicitly being taught.
4. The persistence to continue working on a problem or issue past the point where most people would give up.

These are all strengths that I've found to be a valuable part of my own science training and have helped me greatly in my career. But I've also found that these strengths can present challenges in the private-sector environment, and we need to know when to use them and when to use a different approach.

When overused, these strengths result in the PhD stereotypes that cause the biggest problems in an industry environment, both for the PhD and for their managers. If we look into these stereotypes, it will guide us on how to adapt some of our strengths when we move from a research lab to a private-sector environment.

> *I'm pretty sure that for every characteristic that I've considered a strength, there was also a place where it turned out to be a weakness. So, when I think of my strengths, I acknowledge that sometimes they can become my weaknesses, depending on the situation.*
> – Sona Hosseini, PhD in Engineering; Applied Science, Research, and Instrument Scientist at Jet Propulsion Lab[7]

The PhD Stereotypes in Industry

Digging further into my interviews with industry managers, I have identified three specific behaviors that managers see as the most problematic.

Stereotypical behavior #1: They don't focus on what matters

Industry managers report that the PhDs they've worked with often show a lack of focus on work that will deliver the results that the company needs:

> *The employees with PhDs seem to be more interested in chasing an interesting problem until it is fully understood, which is just what they had to do to complete their degree. It's hard to have the discipline to turn*

your back on an interesting problem and focus on the customer's needs instead. In some cases, you may never understand exactly why what you are doing is working. That's a difficult thing for inquisitive minds.
– Tom Baur, MS in Astrophysics, founder, CTO, and chair of the board of Meadowlark Optics [3]

Digging deeper, this lack of focus tends to appear in two specific habits:

– **'Focus' habit #1: PhD scientists are more focused on what they find interesting than what the customer wants.** Some non-scientists in industry think PhDs are life-long academic types who just want to study interesting problems for the rest of their careers. They expect that PhD scientists are likely to explore some interesting tangent rather than work on the problems that will result in profit for the company:

I've had too many problems with PhDs who spend too much time chasing things they are curious about and aren't focused enough on the problems we have to solve.
– Former CEO in the scientific instrument industry

This concern is understandable, because science research is often open-ended, and scientists know that following a new idea to see where it leads sometimes leads to important new discoveries. For many PhD scientists, the transition to industry requires an abrupt shift from the more open and curious environment they enjoyed as a graduate student:

As a graduate student I just worked long days and it didn't really matter how much time I spent on any specific problem. Suddenly, I was working in a company, and it mattered how much time I spent on each project, because my labor hours were all billed to specific projects and customers.
– Christina C. C. Willis, PhD laser scientist[8]

– **'Focus' habit #2: PhD scientists have a hard time moving on and leaving questions unanswered.** The second aspect of this stereotypical behavior is that we scientists have a tendency to continue working on a problem until we understand every little detail. This behavior probably comes as a natural result of a scientist's curious nature, but our postgraduate training also encourages this habit, as discussed in the 'what is pursued' principle in Chapter 2. When we were working on our dissertation projects, we knew very well that we'd better investigate every aspect of our chosen problem, because our thesis committee was going to subject us to very intense and thorough questioning when we defended. The result is that PhD scientists tend to keep working for improved results even if a project already meets the defined goal. This can be frustrating for the manager:

The PhDs I've worked with always want to keep working on the product, developing it and making it better, even after they've met the specification. But the customer won't pay for 'better,' so we don't want to pay the PhD to keep working on it. Why can't they understand when the product is good enough?
– Product manager in the industrial automation and sensing industry

There's a stereotype of PhD scientists by many in industry that they're kind of useless because they get 'analysis paralysis' and can't make any decisions, or else they want to keep making things better and don't know when to stop and move on to something else. And this stereotype is often true because, in academia, you're taught to stick with a problem until you get it 100% solved. In the real world, no one has time for that. Figure out when it's 80% done and then stop working on it!
– Kate Bechtel, PhD in Physical and Analytical Chemistry, Biophotonics Fellow at Triple Ring Technologies[9]

Many industry managers believe that scientists don't understand the difference between product development and science research, and input from the scientists themselves acknowledges that this is true:

One of the main challenges I faced in industry was switching from science research to product development. These two activities are very different. There are many product development concepts that I had not been exposed to in research. For example, I had no idea what a product development cycle was. The concept of product requirements was also new. As a scientist you tend to pursue the absolute best performance you can get out of your equipment. In product development, you have to define requirements that every instrument produced can meet, and this often requires a compromise.
– Antoine Daridon, PhD in Analytical Chemistry, Business Development and Marketing Manager, Metrolab Technology SA[10]

If you put a couple of science PhDs together in a room, they probably won't ever come out with a viable product, but if engineers and scientists work together on a team, you can achieve a great result.
– Tanja Beshear, MS in Physics, Senior Quality Systems Manager at Medtronic [11]

Stereotypical behavior #2: They try to be 'too smart'

As reported by industry managers, this 'too smart' behavior typically manifests in three specific habits.

- **'Too smart' habit #1: They need to be the smartest person in the room.** One of the most common complaints is that PhD scientists seem to have a strong need to show others that they are smart:

  One of the real hang-ups that scientists and engineers have is they feel like they have to be the expert and be able to provide all the answers. One thing you learn very quickly in business is just how much you don't know, and how much you need to rely on other people. The sooner that you learn that, the more successful you will be.
  – Peter S. Fiske, PhD Geological and Environmental Sciences, Executive Director for the National Alliance for Water Innovation [12]

  In an industry team environment, this often shows up as the PhD scientist suggesting that their knowledge and expertise are superior to others on the team, such as the engineers, who have been trained in a specific discipline. This is frustrating to the other team members who may have years of training in their specific discipline and may resent being told a better way to do their job by a newly minted PhD scientist:

  Just because you 'played an engineer on TV' as a grad student does not mean you're a real engineer. You have your expertise, and you have to rely on your colleagues to be good based on their own expertise. You may have an expert in mechanical engineering and another expert in electrical engineering and another expert in software, and you can't be going around telling them how to do their jobs.
  – Kate Bechtel, PhD Physical and Analytical Chemistry, Biophotonics Fellow at Triple Ring Technologies[9]

  I suspect most of us who pursued a graduate degree regard our intelligence as one of our most valuable assets. Academia teaches us that we succeed by being smart and being correct, with the measures of success including passing the test, completing the degree, receiving tenure, or even winning the Nobel Prize. PhDs spend more time in academia than most others on an industry team, and during that time they are judged on their ability to work alone and be the experts on every aspect of their projects. It's natural that they might emerge with the habit of 'being the expert,' but this behavior can limit your performance on a team and can be a significant barrier to a successful industry career.

- **'Too smart' habit #2: They like to find fault with others' ideas.** People in industry are more focused on finding solutions than faults, and people who frequently point out failings are seen as hard to work with:

  My experience is that most scientists can benefit from learning more people skills. As scientists, we need to know what we are talking about but not come across as arrogant know-it-alls. Generally speaking, people hire folks they want to work with. You've got to be able to convince people that you are not threatening and that you are willing to learn from them. Part of this is to be willing to discuss your ideas with people who may know little about your expertise but are very skilled at what they do.
  – Jason Ensher, PhD in Physics, Executive Vice President and Chief Technology Officer at Insight Photonic Solutions [13]

  This behavior likely results from an important aspect of the scientific method. A hypothesis is proposed to explain an observed phenomenon, and then everyone tests the hell out of it to see if they can break it. If no one can find fault, the hypothesis survives. This tends to train us scientists to value finding fault and pointing out why something won't work.

- **'Too smart' habit #3: They tend to speak in a very complex and specific manner.** Industry managers often complain that PhD scientists are given to overly complex descriptions in an attempt to be as precise and accurate as possible, but in doing so, lose the ability to communicate well with others on their team. As scientists we are taught to be very specific in both our written and verbal communication. I suspect that this habit is a result of being trained in an environment where everyone else is prepared to find fault with what we say, and so we learn the habit of using complex and nuanced descriptions in an effort to avoid being wrong. While understandable, this habit can lead to complex descriptions that can be difficult for people less familiar with the details to understand.

  Industry teams typically contain people with a diverse range of education and backgrounds, and so straightforward descriptions are much more effective, even if they do not explain every detail. Complex descriptions may make you seem very intelligent, but they are not very effective.

  Get to the point. As scientists, we are trained to describe a ton of details and background information before we give the final results. This is the very nature of how graduate students are trained to write their theses and dissertations. It is how scientists deliver presentations at conferences. For the public or policymakers this approach basically

needs to be flipped. The key points or findings need to be delivered very early and it needs to be concise (think elevator speech).
– *Marshall Shepherd*[14]

It is interesting to note that this habit shows a focus on understanding over results, which is the key academia/industry difference listed in the 'what is pursued' principle in Chapter 2. In communication, it is most important that the person we are speaking with receives the main message, not that we make a statement that is perfectly accurate and cannot be disputed.

Simplicity is the essence of effective communication. If you don't grasp the importance of it, you don't grasp effective communication.

– Randy Olson in *Houston, We Have a Narrative*[15]

Stereotypical behavior #3: They are slow to decide

The third major complaint is that PhDs frequently want to continue studying a problem and do more analysis, to be absolutely certain before making a decision or a recommendation. But a private-sector project schedule typically does not have the time for such thorough analysis:

Scientists come from a culture where technical accuracy is absolutely paramount. It completely frames all discussion. When they move into an industry environment, they see teams implementing imperfect technical solutions and making imperfect choices, and they simply don't understand. Their natural tendency is to say the problem needs more study, but this is often impractical. The whole world of business is about tradeoffs and the time value of making a decision now versus later. The fact is, at a certain point you have to make a decision based on incomplete data and move on.
– Peter S. Fiske, PhD in Geological and Environmental Sciences, Executive Director for the National Alliance for Water Innovation [12]

This behavior is also understandable, as the process of publishing peer-reviewed research encourages academic researchers to hold off on submitting results until they are very certain it is accurate. This practice is further reinforced in the PhD dissertation defense, as no candidate would stand up to defend his thesis without being absolutely sure one could justify the results.

But technical companies in the private sector are about taking science developed in a lab and selling it to customers outside of the controlled environment of a laboratory. Customer demand is never certain. Vendors who are needed to supply critical materials can run into business problems of their own; there are uncertain macroeconomic conditions to consider. In addition, a company is a collection of human beings with their own private lives, each with variables that cannot be predicted. This involves so many uncertainties that a "right answer" rarely if ever exists. In environment where there is no right answer, there's a limit to the amount of data collection and analysis that is productive. Product development teams frequently have to move forward never knowing for sure if the path they choose is right.

As scientists we learn that in the lab results can be predicted if the proper models are used and enough information is collected. But in the world of product development, predicting outcomes is not nearly as successful; and experienced business leaders learn that there quickly comes a point when further analysis is not productive, and action must be taken. This is not always an easy lesson for the early-career PhD scientist.

An advanced degree acts as a red flag that you are 'caught up in your head,' you are 'overly cerebral,' and you overthink things.

– Randy Olson in *Houston, We Have a Narrative*[15]

Don't Be That PhD

Certainly, there are many PhD scientists do not fit these stereotypes. But from the feedback I've collected from industry managers, the behaviors are common enough to produce a hesitancy to hire PhDs.

Don't be that PhD. Focus on developing habits that bring productivity in the private sector. Develop a playbook that will help you win the game!

Playbook story: 90% of PhDs

While I was developing the content for my private-sector career workshops a few years ago, I reached out to many of the industry managers I know to get their input on how PhD scientists in their companies have performed—what they've done well and what they haven't done so well.

One manager I contacted had been the CEO of a company that I'd worked at many years earlier. We hadn't seen each other in about 7 years, so I contacted him to let him know what I was doing with TurningScience and to learn from his experience working with PhD scientists and engineers as a high-tech company executive.

I told him I was developing a variety of courses to teach PhD scientists how to transition effectively into the private sector, and that his input would be helpful. I asked him, "What has been your experience with PhDs in the companies that you have led?" He looked straight at me and said, "Dave, in my experience, 90% of PhDs are worthless in industry!" Then he paused for a minute and added, "Well, you were one of the 10%. That's why we hired you."

"Gee, thanks," I added. "That makes me feel much better. But seriously, that's a very large percentage who are not effective. What are the problems or behaviors that lead them to be so non-productive?"

His response: "Three specific things come to mind. First, they always think they're the smartest one in the room. Second, they're always finding fault with other team members' ideas. That just doesn't work in a team environment. Third, they don't ever want to make a decision. They always want to collect more data and do more analysis. We don't have time for that."

Rarely have I received feedback as direct as that. The first point grabbed me right away. 'I'm a scientist!' I thought to myself. 'Of course, I'm smart! That is one of my most valuable strengths! Can't I be proud of my strengths? Doesn't a bodybuilder usually think they are the strongest person in the room? And doesn't everyone generally agree? Why can't I be the smartest one in the room?

'And finding fault with others' ideas is actually part of the scientific method! When a scientist presents a hypothesis, other scientists join in to test it, effectively trying their best to disprove the hypothesis. Attempting to find fault with others' ideas is how science moves forward. If I argue against a point you make, I'm not being difficult, I'm simply being a good fellow scientist aiding you in the search for truth.

'And as for being slow to make decisions, isn't that what we scientists are trained to do? We spend years collecting data and performing careful analysis to ensure our theses are correct. Certainty is very important in science research. I would never have been allowed to graduate with my PhD based on a "best guess." How can we be blamed for wanting to be accurate?'

But as quick as I was to defend these habits, I'd also had enough private-sector experience to realize that the behaviors he mentioned are in fact counterproductive in the fast-paced and team-oriented setting of industry. There is a difference between being smart and showing everyone how smart you are. If you always feel the need to be right, you're probably much less likely to listen to the others on your team.

While it's important to identify bad ideas in industry, a product-development team is generally more interested in finding what will work than pointing out what won't. It's more effective to suggest a test that will make the point with data than act like you have all the answers already.

Finally, it is absolutely true that decisions must be made quickly in industry, and there is often not enough time to collect enough data for absolute certainty. In fact, I've learned that most of the time there are important questions that just can't be answered with data. Too many things that matter in product development, such as customer behavior and macroeconomic conditions, simply cannot be predicted. Thinking that more data or more analysis will help us find the right answer is procrastination.

After a bit of reflection, it was clear to me that he was right. I later came to realize that in 20 seconds he had nailed all three of the key points that were present in the input I'd receive from other industry managers, but none of the others were as succinct or as direct.

Interview excerpt: Kate Bechtel on the PhD Stereotype[9]

Kate has a PhD in Physical & Analytical Chemistry from Stanford and is a Biophotonics Fellow at Triple Ring Technologies in the San Francisco Bay area. Her full bio can be found in the Interviewee Bio section.

Dave: Have you seen other PhD scientist habits that don't work well in industry?

Kate: There's a stereotype of PhD scientists by many in industry that they're kind of useless because they get 'analysis paralysis' and can't make any decisions, or else they want to keep making things better and don't know when to stop and move on to something else.

And this stereotype is often true because, in academia, you're taught to stick with a problem until you get it 100% solved. Then you document the path you took to get from point A to point Z in extreme detail and make sure you understand every detail. In the real world, no one has time for that. You just need it to be good enough. Figure out when it's 80% done, and then stop working on it!

Dave: I was absolutely that PhD physicist who just kept working on the problem to keep making it better. The transition to a product-development environment was a real challenge. There were so many times where I realized I wasn't focusing on what was important, and I had to keep reminding myself to think differently.

Kate: I had those same habits, and it took me time to make this transition as well. I'm eternally grateful to my bosses, the founders of the company, who put up with me and took the time to train me, because I was not very effective when I first started. In fact, I remember early on I was in our office kitchen, and I was explaining to the CEO of the company how a particular project was going. I was outlining the details of every single approach I was considering, giving him the pros and cons of each option. He finally got fed up and said, "Just pick a direction and go with it! You need to move forward!" That really stuck with me.

Dave: We've learned that making a mistake might be fatal, such as when we are standing up to defend our PhD thesis. But that's a different kind of 'fail.'

Kate: Yes! I would say that in most things it's really okay to be wrong, and it's okay to fail. In fact, an example of a very effective staff member is someone who makes a mistake and then very quickly says, "Oh, I was wrong. We need to try a different approach." No one has any ill will toward that person. No one says, "I can't believe you made a mistake," because that's how you make progress.

Interview excerpt: Kerstin Schierle–Arndt on helping scientists transition to industry[16]

Kerstin is a PhD chemist and Vice President of Research Inorganic Materials and Synthesis with BASF. Her full bio can be found in the Interviewee Bio section.

Dave: Many scientists find it challenging to transition from university, where the goal is creating knowledge, to the private sector, where the goal is creating profit. How was that transition for you?

Kerstin: Evry transition has some challenges but that's what makes it interesting. I was not so threatened by going to a place where it's all about profit because I find that very practical. If you develop good products, or if you improve existing products, then society is better off than it was before. Profit is also how you create jobs, so I don't see anything wrong with that.

The way work gets done in industry is very, very different and that was maybe the most difficult part of the transition for me. Technicians do most of the hands-on work because they are much more experienced handling chemicals on a large scale than a chemist who is just coming from university does. In this environment, they are the hands and the eyes, and they have good ideas. Then you as the researcher analyze and digest the results and decide on the path forward. It's quite different than when you are in university doing experiments yourself, and it was a challenge to adjust.

Another big difference is that the teams you work on are much bigger than when you are in university. There are people responsible for marketing and production in addition to the chemists who are leading the research. You have more people available to think about the right approach and ask the right questions, but you also have to manage the communication between the different specializations, and that is very different than in university. This approach of having all the important functions in one group makes you much more efficient, but I remember it being very exhausting at first. You get used to it in time.

Dave: I've seen scientists struggle when they have to go from being the project expert to being only one voice on a team of experts. How does BASF help new scientists manage this transition?

Kerstin: BASF has a lot of training programs for beginners. There is a whole curriculum for people during their first year at BASF. They have courses that cover financial basics and how to lead technicians and many other topics that scientists don't generally learn about in university. New employees are assigned an advisor to help them as well, so you get all the tools you need. People I know who work at other large companies have similar programs, but I don't think that's possible for a startup or even a smaller company.

Of course, there are other things that matter such as having the right mindset and being a team player. I have the impression that in most cases people who are hired at BASF somehow just fit the team well. It might have something to do with the people who apply to BASF, that perhaps they like how we work and so they fit in, but we also have a strong focus on the interviewing process. When we hire people at BASF there are long interview rounds that last at least one day. During the interviews, people will meet with their future boss, the other research team members, and with people in the business roles as well. We tend to get a pretty good feeling whether someone will fit into the team or not.

Then when a new hire starts, they are assigned to a good introductory project. We have a system of mentors, and we also have a system where people share offices. A new hire will typically share an office with someone who is more experienced, and it helps them to get started. It all works quite well.

THE PATH TO SELF-
IMPROVEMENT IS RARELY A
COMFORTABLE STROLL
TurningScience

Chapter 4
Your Private-Sector Playbook

Habits that Successful Industry Scientists Learn Quickly

The previous chapter described some of the stereotypical scientist behaviors that are not very helpful in an industry environment. Of course, many scientists have built very successful careers in the private sector, but they are successful primarily because they have learned quickly and adapted to the private-sector environment. Most of us do not learn much if anything about the private sector as graduate students, and so few of us enter the private sector with these habits.[1]

Some of us have been lucky enough to have patient managers to mentor us and help us develop new habits. I was one of these fortunate PhDs, but many are not so lucky. Not all managers are good mentors. In addition, many technical companies have managers with business or marketing backgrounds, so they may not understand how to transition a PhD scientist into a product development environment.

This chapter will help you learn how to be more productive and more effective in the private sector. Here are five critical habits that scientists who are successful in the private sector practice regularly.

Habit 1: Help the Company Make Money

The very first principle listed in Chapter 2 describes that 'what is created' in industry is profit. This is the most fundamental of all the principles and so it forms the basis of the first habit. A scientist who is successful maintains a consistent focus on how the work helps the company make money and be profitable.

During our time in university most of us had the freedom to work on an interesting problem simply because it was novel and represented an advance in knowledge. After all, it might lead to a new publication, and isn't that the primary metric of success in academia? As a graduate student making a meager stipend, our time didn't cost much anyway. Once we have jobs in the real world making a real salary, the value of our time increases dramatically. Many scientists are surprised to realize just how much this changes:

> *A really big difference (moving to industry) was recognizing the cost of my time. The cost–benefit analysis of how you spend your time changes dramatically between graduate school and an industry job. As a graduate student, you might be asked to spend several days building a piece of equipment instead of spending the money to buy it, because your time was cheap. As a company employee, it's generally the opposite: it's usually better to just buy the part you need so you can spend your time on something more important than digging around the lab to find spare parts or trying to make it yourself.*
>
> – Christina C. C. Willis, PhD laser scientist[2]

In the private sector, the cost of your time is more than just your salary. Your cost to your employer includes additional costs that they must pay, such as taxes, benefits, supplies you use, and even the floor space for your cubical or office. If you are working on a contract for the government or another company, they may be directly billed for your time at a higher rate that includes many other costs as well as some margin factored in for profit. Much of the time, overhead rates are in the 25–50% range, but in some industries they can be over 200%,[3,4] meaning that your billed price is more than three times your salary!

When you are working for a company, if it doesn't ultimately add to the company's bottom line in some way, it's not worth doing. Don't get lost in work that is fun but that is not likely to result in real value. Make sure your work will reduce costs, increase yields, or result in a new product with a clearly identified customer. The scientist who can identify exactly how their efforts help the company be more profitable is a far more valuable employee, and their job will be much more secure.

Habit 2: Figure Out What Matters and What Doesn't

As described in the 'what is pursued' and 'what is rewarded' principles in Chapter 2, the private sector is about getting results quickly and efficiently. Therefore, an effective team figures out what is necessary to get to the result and what is unnecessary:

> *(When) you are building a product… there are things that are important and have to be done right now. Other things may seem important but don't really matter. It's critical to be able to assess a lot of pressures, and then determine which ones are critical and which will not really matter in the end.*
>
> – Ashok Balakrishnan, PhD in Physics, CTO and co-CEO of Enablence Technologies[5]

Author's note: Making money for your company

Here is a list of potential tasks a scientist might consider when working in a technology company. These are ranked from 'High Value/Low Risk' to 'Low Value/High Risk,' based on a) Value: the likelihood the task will help the company make money, and b) Risk: the likelihood that your manager becomes concerned about your productivity if they observe you working on this task.

High Value / Low Risk

- Developing a new technology for an application that solves a clearly defined problem
- Developing an improvement to an existing product with paying customers
- Reducing the cost of an existing product with paying customers
- Working to increase the final test yield of a product so fewer parts are scrapped or reworked
- Studying a failure mode that is limiting the useful life of a product
- Rewriting a product user manual to correct an error that has caused customers to damage the product
- Any project that is solving a problem that is costing the company money

Possible Value / Medium Risk

- Developing an exciting new technology that might someday result in a successful product
- Working for a week to demonstrate 'hero' results because there might be customers interested in paying for higher performance
- Writing an invention disclosure on a new idea you had so that you might get a patent in your name
- Rewriting a product user manual to have a more attractive layout

Low Value / High Risk

- Working out a derivation that your colleague completed so you can understand the principle as well as they do
- Traveling to a conference to present a paper just to get another publication on your resume/CV
- Spending a week to cover up a mistake simply so you don't have to tell your boss
- Machining a part for your project yourself rather than asking a technician to do it because you enjoy working in the shop

Making this shift can be a challenge for scientists who are used to a more open investigative approach to their work:

> *When a PhD comes out of an academic lab, they're in this really exploratory mindset. When you're trying to build a product, you generally have to be a lot more focused.*
> – Marinna Madrid, PhD in Physics, co-founder of Cellino Biotech[6]

Guidelines for figuring out what matters

Figuring out which tasks are critical and which are not can be challenging to do, but it does get easier with experience. I've found three guidelines to be very helpful when I'm trying to determine what matters and what doesn't matter.

Guideline #1: What Are the Requirements?

The best place to start is the formal requirements for the project. If you are working on a product or a deliverable for a paying customer, there should be a contract or written specifications that clearly outline what needs to be done. If your project is internal to the company, there should be a written scope or objective that outlines specifically what is to be done. Smart, creative scientists need to resist the natural urge to make assumptions about what might be involved based on a cursory understanding and stay focused specifically on 'the problem at hand:'

> *It's also very important (in industry) to focus on the problem at hand and not waste time on non-critical details.... Once I jumped into business and realized the problem at hand was that a paying customer expected me to deliver something to them, it wasn't so hard to make that transition (from academia).*
> – Chris Myatt, PhD in Physics, founder and CEO of LightDeck Diagnostics[7]

A concern that I hear frequently from industry managers is that their PhD scientists feel compelled to keep working on a project to make it 'better and better' with little regard for the specifications. Customers pay for performance that meets specifications, and not for performance beyond the specifications. If a customer truly needs higher performance, the company needs to consider a product variant with higher performance and a corresponding higher price. Work that is spent on exceeding the specifications beyond the intended confidence margin is a waste of time and money.

Ultimately, figuring out what matters and what doesn't is about understanding what is important and what is not from the perspective of the entire project, and an important aspect of this is personal time management on the part of each team member. In industry you are frequently faced with many pressures all at once. You have to learn to identify which tasks will end up making a difference and which ones will not. Otherwise, you will be consumed with too much work to do and too little progress to show for it.

Guideline #2: Results or Understanding?

The second guideline follows directly from the 'what is pursued' principle in Chapter 2, but our curious nature as scientists means that regular reminders are prudent to help us focus on what matters. When you are considering the importance of a task, think through specifically how it will get you to the required results. Can you quantify exactly how completing the task will move the team toward the goal, or will it merely provide interesting information?

When I was working for the company that built laser-based combustion monitoring systems, I was once faced with a problem that involved uncertainty in the optical loss when fusing two different types of optical fiber together. This uncertainty was causing problems for the technicians who were installing our products on the customers' furnaces. We found that were not able to accurately predict the loss of a splice based on the information the fusion splicer gave us after the splice. My scientist brain kicked in, and I became excited to think of the full splice characterization program we could put together. With the right experiments, we could do our own full characterization of the splices we needed to do, such that we could predict the splice loss much better. The problem is that this would take weeks to complete. I had to put my manager's hat on and consider whether that would be time well-invested. By applying the 'results or understanding' guideline, I realized that all we really needed to know was whether the splice loss was below the value that would allow the system to work, or if it was above this value. If it was too lossy, it didn't matter how lossy, because the technician was going to break the splice and redo it anyway. All that mattered was that we had enough information to decide if it was a pass or fail. That would be sufficient to give us the results we wanted. So, by applying this guideline, a full characterization project that would have taken six weeks was revised to a pass/fail characterization that took only two weeks.

Guideline #3: Do a 'Gut Check'

This guideline is related to personal motivations for completing a task and may pertain more to personal tasks than to tasks on a team project. The intent is to think about the task you are considering and be aware of your emotions related to the task. Are you excited to do it? That may indicate that your motivation is more personal than practical. Are you feeling tense if you don't complete the task? That may mean the task is more about making you comfortable than meeting the requirements, suggesting again that the motivation may be more personal than practical.

> *In this environment, you learn what things really matter and what things don't. I'll give you an example that was very helpful for me. When I*

needed to send an email about something important, it used to take me days to think about the best way to get my point across. Then I would be so nervous wondering what they would think and how they would respond. I realized that in the startup world I couldn't afford a few days to think about every email, so I decided to take only a few seconds and do my best and see what happened. Surprisingly, I found that most people don't really care exactly what you put in an email. If they like you, all the stuff I was nervous about is completely irrelevant. What matters most is to send the email on time.

– Yasaman Soudagar, PhD in Physics, co-founder and CEO of Neurescence[8]

Does the task we are considering perhaps just reduce the risk on a concern that was not high risk to begin with? Good managers have tools for risk identification and ranking,[9] and low-risk items are not dealt with now, as they only distract resources from the critical functions. The bottom line is that if you are working on something that addresses a concern or question that is not high risk, there is probably something else you ought to be working on instead.

When we do this 'gut check,' we may find that focusing on what really matters is particularly challenging when we were hoping for a different outcome, based on personal preferences or a specific technical idea that we liked. But if we want to be effective, it is important to be honest with ourselves:

Being able to face the truth – whether it's something you like or not – is important.... Sometimes you are in the middle of developing a product and then you realize, 'Uh oh, this is not the way to go.' You have to be honest with yourself and change direction quickly if you want to be successful, regardless of what you prefer.... You have to deliver value, so be honest what that value is.

– Ashok Balakrishnan, PhD in Physics, CTO and co-CEO of Enablence Technologies[5]

Habit 3: Be Effective, Not Smart

That's right, I said stop trying to be so smart! If you want to be successful in industry, you need to focus on getting things done. No one cares how smart you are if you aren't helping the company get the results it needs as quickly as possible.

This habit directly addresses the integration of three of the game rules in the last chapter:

> **Author's note: Freedom and creativity in industry**
>
> Some industry scientists feel that the need to focus on what a customer will buy limits their freedom and creativity, but this is because they are focusing on the wrong issues. Certainly, they no longer have the freedom to work on any interesting topic that might be novel and someday result in a publication, but this doesn't mean they have lost their freedom. It is important that they focus on work that gets to the results that their company needs—but the methods by which they achieve those results, and the specific path to get there, are often completely up to them. This is where freedom and creativity are expressed in the private sector, in the creative solutions to the problems that they face. The manager is likely too busy to have time or the interest to specify exactly how a project is completed. They are looking for someone with the creativity and the initiative of a PhD scientist to figure that out. That's why they hired a PhD scientist in the first place!
>
> While it may seem like the loss of academic freedom is a negative, most people find that the focus helps their productivity. Open-ended freedom sounds like a great thing, but it can result in undirected wandering in pursuit of things that lead nowhere. A paying customer does wonders to focus the effort of a group, so while in principle there is a reduction of freedom, there is generally a significant increase in the efficiency and the effectiveness of the work that is done. Freedom to pursue whatever research path fancies you sounds like a great thing, but often results in publications that only a handful of people will ever read. When a customer asks you to develop a challenging new solution to some big problems they face, they are to some degree defining what you will work on, but you can be pretty sure that your work will make a difference within only a year or two, and you get to define specifically how to solve a problem. That is exciting!

What is pursued: Results
What is rewarded: Speed
How one succeeds: Interaction

The wording of this habit 'be effective, not smart' is meant to be a bit amusing, or 'tongue-in-cheek,' but it is a good reminder for the PhD scientist to focus on achieving the desired outcome rather than on proving they are smart or correct. This habit is perhaps one of the hardest for a scientist to adopt, given the tendency to prove we are smart that most of us learned in an academic environment and the fact that many of us build our self-image around our intelligence.

Principles that remind me not to be too smart

I've developed three principles that help me maintain the habit.

Principle #1: If you get results quickly, people will assume you are smart

In the private sector, people who know how to get things done quickly are highly valuable and highly respected, much the same way intelligence is respected in academia. The value is not in having the right answer, it's in finding the right answer quickly, i.e., if you don't know the right answer, find someone who does.

> *One very important thing for me has been learning to work with and leverage the skills of other people and not let a fear of sounding stupid keep me from asking for their help.*
> – Jason Ensher, PhD in Physics, Executive Vice President and Chief Technology Officer at Insight Photonic Solutions[10]

I tell people who work on my team, "I don't care if you don't know the answer to a problem we are struggling with. What I care about is whether you can find the answer quickly. If that means you need to ask someone else in the company, or even hire a consultant, that's much better than spending a week or a month trying to become the expert yourself."

Principle #2: Winning as a team is more fun

In the results-driven world of the private sector, speed is essential. Companies address this by building teams of people with diverse skill sets who work together to solve problems quickly. This approach is much faster than for one person to try to become expert in everything that is required to succeed, but it can also be more fun to be part of a group that is working hard to achieve something new.

The satisfaction that comes from being proven correct or from knowing that you are smart is a private celebration. When you win as a team, everyone gets to join in the celebration, and everyone gets that great feeling. Rather than searching for things that give you that dopamine hit associated with a personal victory, look for things that allow you to enjoy the team victory. This might include any of the following:

- Recognizing when a team member has a great idea that will help you all win.
- Dividing a critical task into parts so that several team members can contribute.
- Thanking a team member for pointing out a mistake you made and helping you all move forward.

These actions may be a bit challenging for PhD scientists to embrace at first, but the habit of being effective rather than smart is reinforced as they come to appreciate the great feeling and camaraderie that comes with a team victory.

Principle #3: Everyone has something valuable to teach

Industry teams typically comprise a wide range of education levels and work experience, often spanning from recent graduates with advanced degrees in science or business to technicians with only a high school degree and 25 years of experience in a manufacturing environment. I've found that I can usually learn something valuable from every one of these people. An attitude that has served me very well over my career is to approach each coworker as someone who can teach me something valuable.

An unfortunate habit of some PhDs is to look down on those with less training as less valuable members of the team. My experience has shown me that people with less training are valuable, and often more reliably productive, members of the team. A good technician is extremely valuable, because their attention to detail and tolerance for repetition are critical for manufacturing a product that consistently meets performance and reliability standards. An experienced technician has been working in a manufacturing environment for a long time, and they usually know it very well. A scientist usually has a very different set of experiences, and can learn a lot from an experienced technician. I've heard experienced managers say that a good technician is worth their weight in gold, and I agree.

Design engineers are occasionally looked down upon by PhD scientists as their less-educated technical cousins, because most have only an undergraduate degree. But their deep level of understanding of their discipline is a critical aspect of product design and can provide useful insight to the scientist:

> *I've found that physicists tend to develop a strong, overarching perspective on the product – to understand the whole picture, but they can learn a lot about the details from the design engineers.*
> – Jason Ensher, PhD in Physics, Executive Vice President and Chief Technology Officer at Insight Photonic Solutions[10]

Many scientists feel disdain for the entire concept of sales, and so did I at one point. It's natural to think of the sleazy 'used-car salesman' as the quintessential salesperson, but I've found that this is far from reality. From the sales managers I have worked with, I have learned so much about communicating with customers and even more about influence and persuasion. I've learned to determine what factors matter most to a customer, and this skill has benefited me greatly in my private life as well.

The people who work in human resources have reminded me that my coworkers are human beings, and I need to think of their concerns beyond the work that we are doing if I am to understand what motivates them and what limits them.

Conversations with company executives have opened my eyes to the world of business, something I never imagined I'd have any interest in when I was a graduate student. Some of my most memorable and valuable conversations have been with the CEO over a couple of beers late into the evening while on a business trip. Business is how the world functions, and the scientist who embraces this fact will be more effective.

The successful scientist in industry learns how to be effective rather than smart, and they quickly realize that teamwork is the way to do this. And I mean true teamwork, where everyone plays a critical role in the success of the team, not just the environment most of us experienced in academia where several people with similar expertise worked alongside each other on their own personal projects.

> *The smartest person in the room, I've learned, is usually the person who knows how to tap into the intelligence of every person in the room.*
>
> – Scott Kelly, Engineer, test pilot, and retired astronaut[11]

Interview excerpt: Kate Bechtel on being the smartest[12]

Kate has a PhD in Physical & Analytical Chemistry from Stanford and is a Biophotonics Fellow at Triple Ring Technologies in the San Francisco Bay area. Her full bio can be found in the Interviewee Bio section at the end of this book

Kate: (The transition to industry has) been a difficult process for me to change my behavior and develop an attitude of teamwork and cooperation. I had to learn to value the skills that other people bring and not try to be the smartest person in the room. I'm still working on that.

Dave: That habit of needing to be the smartest person in the room is common among PhDs. I'm still working on that myself.

From our time in academia, we scientists learn that being smart is the primary value. But in the private sector, the primary value is in being effective and getting things done. At the end of the day, getting a finished product out the door is what keeps the company doors open, not how smart you are.

Kate: You have just totally nailed it. In industry, you're part of a team that is collectively trying to solve a problem, and the problem is more important than the egos of the people on the team.

I think of the scene in the Tom Hanks movie *Apollo 13*, where the team members all go into a room and dump all the parts they have on the spacecraft out on the table and say, 'We have to make a scrubber out of these parts.' Everyone just dives in, and it doesn't matter who they are or what their role is, they've just got to solve this problem. Everyone is equally important. That's the attitude that you need in industry.

Dave: I've visited science departments that blend basic and applied research, and they seem to be more in touch with the world outside of academia. The scientists in these departments seem less likely to have the attitude that a PhD makes them better and smarter than the rest of the team. I'd love to see more of that attitude in academia.

Kate: I would, too, and I think you've highlighted the thing that really gets me about academia. It's the elitist attitude that is so common. It's the people who think that they know everything and don't appreciate that a technician who only has a high school degree but has been doing their job for 20 years probably knows far more about it than they do.

Working with people who have been in industry longer than you and have picked up a tremendous amount of knowledge is such a valuable experience. That's what every PhD student needs.

Dave: I've heard many industry managers complain about that lack of understanding in their PhDs. One manager complained that a PhD on their team thought that, because they spent time making a few parts for their dissertation experiment in the machine shop, they could tell the mechanical engineer on the team a better way to design the product.

Kate: That's right. Just because you 'played an engineer on TV' as a grad student does not mean you're a real engineer. You have your expertise, and you have to rely on your colleagues to be good based on their own expertise. You may have an expert in mechanical engineering and another expert in electrical engineering and another expert in software, and you can't be going around telling them how to do their jobs. They have years of experience, and they know what to do.

Author's note: How do successful people accomplish so much?

In my workshops, I'm frequently asked how an ambitious person manages to get everything on their to-do list done. If a person wants to do big things with their life and career, they tend to collect more and more tasks that they need to accomplish. At some point, it simply becomes too much to manage and leaves some people wondering if they have been too ambitious in their goals.

80/20 Principle

Successful people recognize the value of the 80/20 principle, also known as the Pareto Principle,[13] which is the observation that the majority of the impact of what we do comes from a minority of the effort we provide. The literal interpretation of the principal would say that 80% of the outcome comes from a mere 20% of the effort. In short, most of what we do doesn't matter in the end.

Examples of this tendency are observed in all aspects of life and business. Eighty percent of automobile accidents are caused by around 20% of the drivers. Eighty percent of the problems uses have with a particular software application come from around 20% of the bugs. Twenty percent of your clothes account for around 80% of what you actually wear. And when businesses study their productivity, they often find that 20% of their customers account for around 80% of their profit, 20% of their employees create 80% of the value, and 20% of their employees' work time creates about 80% of their productivity.

The 80/20 principle is not any type of natural law but merely a tendency. This suggests that if we can identify and focus on that vital 20% in our to-do list, we can improve our productivity.

Prioritization of your to-do list is a common time-management suggestion, but simply putting things in order from more to less important does not create any more time in your schedule. The 80/20 principle suggests that much of what we do in that non-vital 80% is wasted effort and probably should not be done at all. Although we probably can't ignore the entire 80% without some negative consequences, only when you discard some of the less important tasks do you gain the extra time you need to accomplish what really matters. This suggestion do discard tasks is not as common and not nearly as easy to do.

The Time Management Matrix
A well-known technique that *does* embrace this discarding technique is the Time Management Matrix put forth by Stephen Covey in his well-known book *The Seven Habits of Highly Effective People*.[14] The approach involves sorting tasks into a 2 x 2 matrix based on whether they are important/not important and urgent/not urgent. One then uses a different tactic to deal with the tasks in each quadrant, as shown in the diagram below.

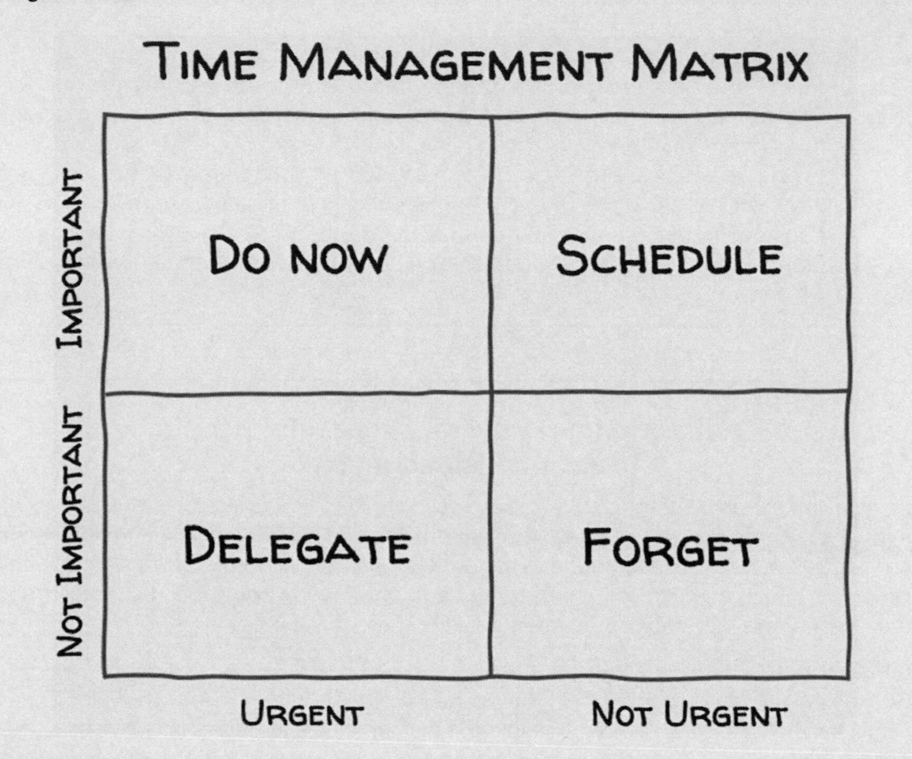

Playbook Sketch 2 The time management matrix

The real value in this technique comes from the discipline of removing everything that is rated 'not important' from your to-do list, thus freeing up time to do the things that are rated 'important.' However, the advice from Covey and many others who have written about the time management matrix focuses on capturing 'distraction' and 'time waste' activities in the lower two 'not important' quadrants. This is valuable for anyone who struggles to avoid inherently unproductive activities. But for those of us who already have the self-discipline to avoid truly time-wasting activities, it doesn't help so much.

The real key to accomplishing your goals in your life and career is in how you define 'important' and 'not important.' Activities that are not a productive use of anyone's time and don't really matter in any significant way certainly belong under 'not important,' but so do activities that are productive and matter only to someone else and their goals.

I suggest that you draw a very clear line between what you consider important and not important, based on the goals and values that are truly important to you. When we look carefully at our to-do lists, most of us find that there are many tasks that other people would like us to do, but they don't really move us toward our own personal goals. These might be things we agreed in some way to do or simply a request from someone we respect, and we feel obliged to help them out. Here is a list of tasks from my recent to-do list that I divided into 'important' and 'not important' based on whether they did or did not fit my own personal goals:

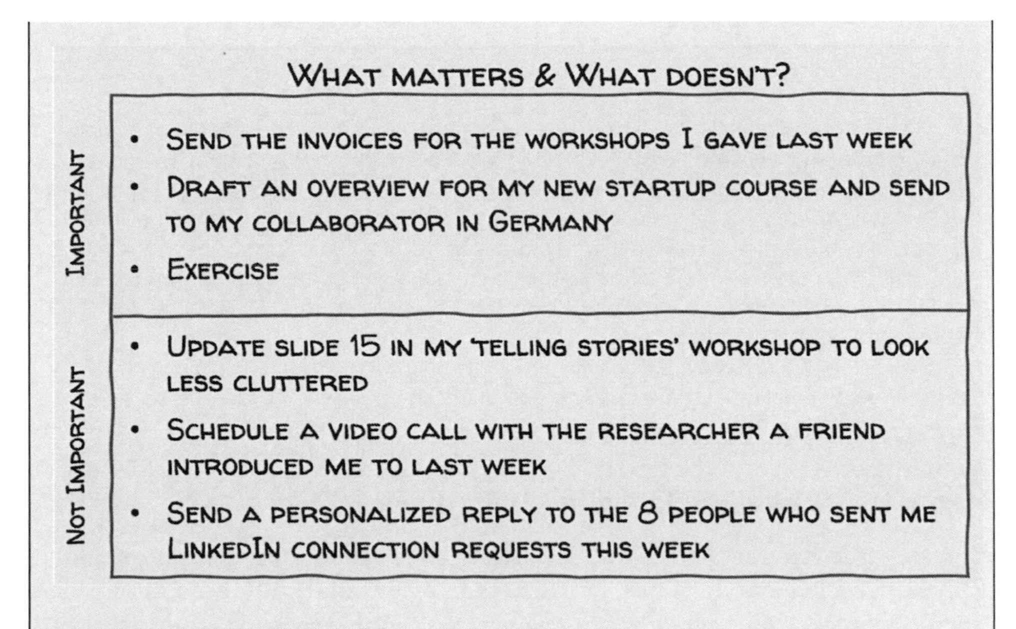

Playbook Sketch 3 Prioritization of some of my tasks.

Relegating some of these to the 'not important' half was difficult. A few of these were related to things I'd agreed to do at one time, and by now I feel I fulfilled my commitment (Very important if you want to make a dent! It's critical that you do what you say you will, but that doesn't mean you are indebted forever.) Others were things I'd like to do to help someone else. Helping others is important, and I suggest you always budget a portion of your time for this, but requests from others can pile up, and if we are not careful, they can steal valuable time from working on what matters to us. Often, we are aware that these tasks are not critical to us, and so we don't do them right away. Instead, we tend to keep them on the list but at the bottom with other 'low priority' tasks that we firmly intend to 'get to someday.' We scientists are a responsible lot, and we can often struggle to decide not to do some of the things that someone asks us to.

The problem with this approach is twofold:

- **Problem 1 -** Leaving these tasks on our list prevents us from telling anyone that we are not planning to do them, and so we often end up just being very late in delivering the results, which does not help our professional image.
- **Problem 2 -** The attention that we give these tasks, even if we only let them regularly distract our attention as we remember that we 'really need to get to them soon,' steals our valuable attention from the things that matter. This is the thing that most damages our ability to accomplish our goals.

The time you need to accomplish what is truly 'important' to you comes from deciding that you *will not do* those things that are neither urgent nor important. I know that this is not easy. One of the reasons this can be hard for us responsible scientist types is that we worry about what bad things will happen if we don't do them. But the whole point is that if we over-prioritize what other people think is important, we are necessarily devaluing what we think is important. If we want to accomplish what is truly important to us, we need to be OK with disappointing others now and then. Not over big really important things, but over things that matter more to them than to us. Tim Ferriss describes this as the "art of letting small bad things happen,'[15] and it is ultimately about figuring out what matters and what doesn't, relative to the results that you want to achieve.

If it won't have a major impact on those important things that you want to accomplish in your career or your life, then it 'won't matter.' Forget it! Don't just keep putting it off, planning to do it someday. The mental energy you expend thinking about it over and over steals your energy from the things you really want to do. Decide that you will not do it, and if appropriate, tell the person who asked you to do it. If they have their own ambitions and are trying to make their own dent in the world, they are far more likely to respect you for setting boundaries than to be disappointed in you. And they will find someone else to help them.

This is the secret to how successful people accomplish so much. It's about what they *don't* allow to waste their time, energy, and attention.

Habit 4: Decide Quickly with Limited Data

Business moves forward when decisions are made and action is taken. Experienced leaders in industry realize that when analyzing a decision, there quickly comes a time when more analysis just leads to delays and lost profit. Analysis is an important foundation for a decision, but you have to recognize when you have reached diminishing returns and it's time to move on.

This can be a challenging habit for scientists to acquire, because we are trained in a field that is all about finding the right answer. We are used to working in an environment where collecting enough data and doing enough analysis ultimately results in a high level of confidence. Many of us who pursue science as a career prefer this type of environment, and we are willing to work as long and hard as it takes to achieve that high degree of confidence. In a world of massive uncertainty, we've chosen a path that lets us try to distill some type of truth out of all the madness. It just requires some intelligence, some training in our chosen field of expertise, and the scientific method. What a wonderful thing.

But this is not the way the private sector works. The private sector is about getting a product to a paying customer to satisfy a need, and so it's about moving quickly and efficiently. This means you don't have all the time you may need to collect all the data and do all the analysis that you might like. A new habit is necessary. That new habit is to decide quickly without all of the data that you might like to have.

Data and analysis are great, and they are a very important part of the process in the private sector. They help to tell you important information about the possible outcomes of each of paths you might be considering. But they won't tell you everything. In the end, a decision has to be made about what path the group will take forward. Speed is important, and as discussed in Chapter 2, much of the time there is not a single right answer.

QUESTIONS WITH A RIGHT ANSWER	QUESTIONS WITH NO RIGHT ANSWER
• WHAT IS THE YIELD STRENGTH OF A SUPPORT FRAME MADE FROM A CUSTOM ALLOY?	• WHEN OUR CUSTOMER RECEIVES OUR QUOTE FOR THE COST OF THE SUPPORT FRAME, WILL THEY DECIDE TO GO WITH ANOTHER SUPPLIER?
• HOW WILL EUROPEAN SHIPPING AND CUSTOMS EXPENSES IMPACT THE PRODUCT COST?	• WHAT WILL HAPPEN TO THE VALUE OF THE EURO IN THE COMING THREE YEARS?
• WHO IS THE MOST EXPERIENCED ELECTRICAL ENGINEER ON THE DESIGN TEAM?	• IF I ASK THE SENIOR ELECTRICAL ENGINEER TO LEAD THIS BIG PROJECT, WILL THEY FIND THE NEW ASSIGNMENT REWARDING, DISAPPOINTING, OR OVERWHELMING?

Playbook Sketch 4 Some questions that have a right answer and some that do not.

I struggled for some time to accept the idea that many decisions do not have a right answer. My trust in the scientific method as a powerful way to find an answer to so many questions led me to suspect that if we just had enough data and access to powerful enough analysis tools, we could, even if only *in principle*, figure out the right answer. This may, in fact, be true. But a problem that has a right answer *in principle* doesn't help us make a decision here and now, when we have a customer waiting on our solution. When you are doing anything in the private sector, there are many things that simply cannot be predicted with data and analysis. There simply is no right answer.

What might be some questions that have no right answer? Consider the comparison shown in Playbook Sketch 4.

The big problem comes when a scientist is faced with having to make a decision or simply a recommendation so that their boss can make a decision, and they struggle to do it with the data they have. This is what leads to the third stereotype listed in the previous chapter—that PhD scientists are slow to decide. The answer is to become more comfortable making that decision when you are hesitant to do so.

How important is the ability to make quick decisions with limited data? Well, that depends on the job you have.

Three types of jobs

In my workshops, I often say that there are three types of jobs. The first type of job is where the employee does what someone tells them to do. This is the type of job where the work product is well-understood, and the company just needs someone to do it. A classic example of the private-sector environment is an assembly technician. The product design and assembly procedures are well-known and documented, and the company just needs someone to sit down and build however many are on the backlog. These jobs are usually paid by the hour, and they are worth a certain amount of money.

Job #1 - Do what someone tells you.
Value = $

The second type of job is where the company needs to answer an important question, and they need somebody to figure out the right answer so that then they can implement the solution. This type of job generally requires a certain level of intelligence as well as training in the skills and knowledge required to figure out the right answer.

Most of us signed up for this second type of job when we decided to be scientists. We wanted to unlock the secrets of the universe. We wanted to figure out the right answer to some aspect of how the universe works. Engineering is a related technical career path that is also this second type of job, because the goal is to design the correct solution to a problem using design rules. There may be more than one right answer, but it is generally expected that at least one right answer does exist.

Because this second type of job requires intelligence and lots of training, it's worth a lot more than the first type of job:

Job #2 - Figure out the right answer.
Value = $$

But there is a third type of job that no university program will train you for. That job involves making decisions when there is no right answer. This is the essence of leadership, and if you aspire to be a leader at any level, you need to be able to make a decision when there is no right answer. If the right answer is obvious, all that is needed is a manager to direct the team according to standard project management principles. Leadership is not required in this case. But most projects in the private sector will run into situations where there is no right answer, and a leader is needed to decide what will be done. This type of job is worth the most money.

Job #3 – Make a decision when there is no right answer.
Value = $$$$

Making decisions is hard, but good leaders make decisions despite this difficulty. Bad leaders often avoid making decisions because they are afraid of making a bad decision. Too much fear of making a mistake can be detrimental, because not making a decision is often even worse than making a bad decision.

Good leaders sometimes make bad decisions that may cause products and companies to fail, but bad decisions can often be reversed if the team is diligent and paying attention. Not making a decision prevents any progress at all. It also sets a bad precedent for the team and builds a culture where indecision is accepted.

Good leaders realize that mistakes are usually not fatal so long as they are diligent at monitoring progress and willing to admit a mistake and change the path if necessary. Good leaders realize that the best way to make fast progress is to take action based on their best assessment of the situation at the time:

> *A very important perspective that I learned from my advisor and use when talking to my team is 'It's OK to make mistakes. We just have to make them faster than our competition.'*
>
> – Roger McGowan, PhD in Physics, Sr. Research Fellow at Boston Scientific Corporation[16]

Three types of labor

The three types of jobs are closely linked to the three types of labor: physical labor, mental labor, and emotional labor. All jobs are made up of some blend of these three types. Few jobs are exclusively one type of labor, but there is a strong correlation. Here's how they match up:

- **Physical labor**: Most closely associated with job type #1, i.e., sweeping a floor, digging a ditch, or building a house. Even being an assembly technician is primarily physical labor, although not hard physical labor. This is the type of job where the employee goes home at the end of the day and leaves their job at the work site.
- **Mental labor:** Most closely associated with job type #2, i.e., solving a complex integral, designing an experiment, or figuring out why the test results this month don't match the results from the previous three months. This is the type of job that one often takes home with them when they are facing a particularly big problem.
- **Emotional labor:** Most closely associated with job type #3, i.e., picking a name for a new product, selecting a new employee from two excellent candidates, or cancelling a big project that is losing money and informing the employees who have spent 2 years working on it. This is the type of job that one often has to learn to not let consume them all day, every day. Some people are simply not cut out for this type of job.

Most of us scientists are quite familiar and comfortable with mental labor, and many of us actively welcome a significant mental challenge. Physical labor may not be as attractive to many of us scientists, but there are many people who love it. Bodybuilders are perhaps the physical labor equivalent of the PhD scientist—someone who loves constantly challenging themselves and truly enjoys that type of labor.

Emotional labor gets far less attention, but it is a critical element of leadership. Emotional labor is required when we do something that is emotionally very challenging, such as making a decision that will impact many people when we don't know for sure what will happen. Author and former business executive Seth Godin speaks about emotional labor frequently in his work and describes how it is more important for our careers now than ever before. He states in his great book *Linchpin: How to Be Indispensable*, "Emotional labor is the task of doing important work, even when it isn't easy,"[17] and "... One of the most difficult types of emotional labor is staring into the abyss of choice and picking a path."[18]

This is worth reflecting on as you develop your career. Are you willing to embrace emotional labor? Are you willing to challenge yourself regularly to train your emotional strength and improve your performance? People who are willing to do emotional labor are not as common, so this is where we can differentiate ourselves and provide a less common value.

> *There is a lot of passive-aggressive behavior in academia. You've got to be much better in industry because there are people who have much better people skills than you do. What I see in the leaders in industry – VP's, senior managers – is something that is hard to put your finger on. You know it when you see it - it's emotional intelligence, emotional control, and self-discipline. It's that ability to control their own reactions but also to be able to read other people. The people who are put in charge are those whose judgment people trust. They are not going to lose their cool. They make data-driven decisions, but they also keep an open mind to how the customer, the business partner, the other engineers see the situation.*
>
> – Jason Ensher, PhD in Physics, Executive Vice President and Chief Technology Officer at Insight Photonic Solutions[10]

Games and emotional labor

Various games involve different types of labor: football takes lots of physical labor, whereas chess takes lots of mental labor. But every game also involves some degree of emotional labor, because one often needs to make decisions when there is no right answer. Games with more uncertainty typically require more emotional labor. For a high-level company leader, developing a technology into a product that meets challenging specifications, attempting to scale it to high volume, and releasing it to an uncertain market requires quite a bit of emotional labor. This is what your manager faces. Can you help them?

As a scientist in the private sector, a critical value you bring to your employer is the technical guidance you provide to management. Your management is looking for guidance from you, as they generally do not have the technical expertise that you do. Can you give them the guidance they need when you have not had the time to completely understand the problem and determine the right answer? Can you take a risk and make a recommendation anyway?

A helpful decision-making technique

Making decisions when there is no right answer is critical to leadership at any level, and it isn't easy, but it brings real value. Practicing this to get better is a good way to develop our strengths for a leadership role.

The best way to get better at making decisions with limited data is to do it. When faced with a decision, make a conscious effort to decide quickly and move on.

So how do good leaders decide quickly without the crippling fear that they made the wrong decision? I can't give you any advice that will make it particularly easy, but I can give you a technique that I found to be very valuable when I'm struggling to make a decision. When I have a decision to make, and there are two or more paths that seem like they might work but I really don't know which to choose, a technique that works very well for me is:

Make a decision...
and then work to make your choice the right decision.

There are two great reasons I like this technique so much:

Reason number one: My ability to predict the future is not nearly as good as I'd like to believe it is. We scientists are used to working on problems that have a right answer and where the framework is deterministic. When we set up an experiment, we are working with the expectation that if we remove enough complexities, we arrive at a simplified system where we can predict the outcome. In the physical sciences, we likely expect to apply a mathematical formula to the phenomenon. Predictability is one of our primary goals!

The problem is that in the 'real world' there are far more factors at play than just the science. This is precisely the challenge of turning science into things people need. We must now take our carefully controlled experiment off the lab bench and turn it into a product or service that can be sold to many customers. So many other things are at play, most of which cannot be predicted at all. A sampling of things that cannot be predicted is provided in Playbook Sketch 5.

THINGS THAT CANNOT BE PREDICTED

- OUR CUSTOMERS' PLANS AND FUTURE ACTIONS
- IF OUR TARGET CUSTOMERS WILL BUY OUR PRODUCT
- CHANGES IN THE MARKETS WE SELL TO
- THE PLANS AND FUTURE ACTIONS OF BUSINESS PARTNERS
- THE DECISIONS OF THE HUMAN BEINGS IN OUR COMPANY
- FUTURE MACROECONOMIC CONDITIONS
- A GLOBAL PANDEMIC

Playbook Sketch 5 Important private-sector factors that cannot be predicted.

No matter how smart we are or how sophisticated the analysis tools we have at our disposal are, we cannot reliably predict the future outside the controlled environment of the science lab. If we have reduced our options to two or three that we think have a reasonable chance of success, and there is no further analysis that will give us the right answer within a few days, we should just pick one and start moving forward.

Continuing to collect more data and further analyze the problem is simply procrastination. We are just wasting time while we fool ourselves into thinking that more time will allow us to predict the future more accurately. The longer we wait to make a decision and take action, the longer it will be since we should have made the decision and taken action.

Reason number two: The second reason I love this technique is that once we make a decision, and we begin to move forward, there are so many things we can do to make that choice successful. I'm certainly not suggesting that success is guaranteed. It's entirely possible that as the future unfolds, the path we chose doesn't work as well as we hoped, and we have to make changes to our plan. However, once we make a decision, we can get the entire team working in the same direction all working to make our choice a success.

Most importantly, once we start moving forward, the problems that we encounter are real problems that we can take actions to resolve, rather than hypothetical problems that we only anticipate but can do nothing about. I don't know how to solve hypothetical problems, and as the U.S. musician Tom Petty once said, "Most things I worry about never happen anyway."[19] Rather than trying to deal with many hypothetical problems, most of which probably will not occur anyway, making a choice and moving forward allows us to see what problems *actually* occur. I do know how to solve *real* problems. That's why I have a team of smart and skilled people.

Once you come to grips with the fact that analysis, modeling, and prediction work far better in a controlled environment of the science lab and they do in the real world of business, the technique of making a choice and working to make it the right decision makes so much more sense.

Interview excerpt: Brit Berry–Pusey on making decisions[20]

Brit has a PhD in biomedical physics from UCLA and is the co-founder and COO of Avenda Health in Santa Monica, California. Her full bio can be found in the Interviewee Bio section.

Dave: I frequently see scientists in industry struggle to make a decision or recommendation when they'd like to have more time and more data. That's emotional labor, not intellectual labor, and that also requires a high EQ.[21]

Brit: I think by nature scientists tend to avoid absolute statements and hesitate to rule anything out because nothing's ever for sure. But when an executive is looking for guidance, a scientist who refuses to give an estimate or commit to a number will really annoy them. I don't think scientists should ever state things as fact that aren't factual, but you need to be able to say something like, 'In my opinion, there's an 80% chance that this is going to work,' and you need to clearly state what the assumptions and the risks are.

At the end of the day, the people you work for are looking for your guidance. They're trying to make a decision about where to take the company, and it is your job to help them. If you can't provide them with information in a way that helps them make their decision, then why are you there? It's important to step out of your scientist shoes and think, 'What is my goal here? Who needs my help and what is their goal? How can I communicate effectively to help us all move forward?'

Habit 5: Persuade Others to Follow You

The final suggestion for the industry playbook is to learn to persuade others that your ideas have merit. If you see the right direction to go, don't wait for others to see it too. Stand up and sell it!

We are always selling something, and the sooner one accepts this fact, the better.

– Yasaman Soudagar, PhD in Physics, co-founder and CEO of Neurescence[8]

In academia, we scientists rarely have to persuade others to follow our ideas. In science, the data speaks for itself. If we want to make a convincing argument in science, we collect enough data and perform enough analysis that

Interview excerpt: Scott Sternberg on decision making in business[22]

Scott has an MS in physics and is the Executive Director of the Boulder Economic Council in Boulder, Colorado. At the time of our interview, he was the Executive Vice President of Services at Vaisala. His full bio can be found in the Interviewee Bio section.

Dave: What was most challenging about making the transition to industry?

Scott: The biggest challenge for me was the change in culture. In science research I was working in a very structured and ultra-precise environment where everything is calculable from first principles, expect perhaps working hours. The business community culture is very different. This is even true in business finance where you might think things would be well defined. Instead, there is a whole culture that exists around interpreting the numbers, even with GAAP, a standard that you might think is an iron-clad framework.

It has taken me years to just frame the question of how business evolves and what rules it follows. Just go to the business section of any bookstore and you'll see many different opinions on how this business organism works.

We talk a lot about fact-based decision-making in business. People often think if they have an infinite number of facts they can make the correct decision. I don't see that in business ever. The more facts you learn, the less you really know what the right decision is. It's the Heisenberg Uncertainty Principle applied to business. There is never a complete set of facts, so decisions in business are often driven by intuition. It was this dilemma that drove me to my definition of brilliance: a unique mix of intelligence, creativity and intuition. Apply that definition to "brilliant" people in history and you get a pretty good working model.

When you look at the iconic CEOs, the people who have really figured out the business community, they have this brilliance. They are smart, well-read, and well-educated, but also highly creative. They try new business models and new ways to drive their employees. At their core, though, they have this intuitive feel of what their customers want – not necessarily what they need, but what they really want. They have a feel for how their employees can actually be adding value to the company every day. Think about what Steve Jobs did at Apple. He embodies all three of those abilities.

the conclusion is self-evident. No scientist submits a paper for publication and then visits the editor in person to make a pitch on why they should publish their paper in the next edition of the journal. They expect that the peer reviewers will see the merit of the work and agree to publish it.

This is not the case in industry. In an environment where there is often no right answer and decisions need to be made with limited data, a scientist cannot trust the data to speak for itself. We need to help others see why we believe it is the best decision or the best course of action.

Persuasion is not always a case of needing to convince people who disagree with you. It might be that no one else has the technical expertise that you do, or it might just be that you know the issue better than anyone else, because that's your job. Business is a game, and everyone is busy playing their own position. They probably haven't given it the thought that you have until you bring it to their attention. This doesn't mean that they will simply listen to anything you suggest, either. They may be uninformed but also skeptical, so you need to make a convincing argument.

Persuasion and influence are important aspects of leadership. One may say, "Do I need to learn persuasion if I don't want to be a manager?" I say, "Yes! Absolutely!" Leadership is not just about being a manager; it's about influencing the outcome of an important decision. If we as scientists say that we want to make a dent in the world, then we are expressing a desire to influence. We may have begun this journey thinking that making a difference was about finding the right answer, but now we've learned that the private sector—the place that gives you the ability to make a big difference—often operates in an environment where there is no right answer. Therefore, the influential scientist knows how to find the right answer when appropriate, and how to make a decision and persuade people to follow that decision when there is no right answer.

The reference that is most often cited for good advice on improving your persuasion skills is the book *Influence: Science and Practice,*[23] by Robert Cialdini, a professor of psychology at the University of Arizona and a world-renowned speaker and consultant on the art of persuasion. Cialdini's book gives an excellent and very thorough discussion of a wide range of methods that are successful at influencing other humans, as well as many stories of how these techniques have been used for both positive and negative outcomes.

Playbook story: Maurice Hilleman and persuasion

When Maurice Hilleman was working to develop a measles vaccine for Merck, he needed to find a reliable source of chickens since the vaccine was created from a weakened virus cultured in chicken eggs. He wanted to ensure the eggs he used were free of a leukemia virus common to chickens so the vaccine would be safe for children. Fortunately, Hilleman finally found a farm in California where the chickens were virus-free, but the breeder did not want to sell him any chickens. Hilleman was ready to leave empty-handed when he noticed that the Director of Research, Dr. Welford Lamoreux, had a familiar accent. He asked Dr. Lamoreux where he was from, and he responded he was from Helena, Montana. Hilleman reached out his hand and said proudly, "Miles City, Montana!" Dr. Lamoreax smiled and replied, "Take them all. A buck apiece."[24]

Persuasion and integrity

When I speak to groups of young scientists, I find they are often resistant to the idea of being persuasive, or 'selling' their ideas. Many of us have unfortunately picked up the false idea that persuasion is about convincing someone to do something that they don't actually want to do, perhaps through trickery or a brute force overpowering of their will. This is not a good view of persuasion.

A much healthier view of persuasion that is fully consistent with the integrity that most scientists hold in high regard is helping another person see the same advantages that you do. If you believe that a particular decision is the best answer for you on the team, and you truly believe you've analyzed it enough that your opinion is valid, what's wrong with convincing the rest of the team that this is the right direction to move? Persuasion in this case is simply helping them see it the way that you do.

> *A really good way to think of persuasion and influence is to think about it like a strategy game, where you have to sell your vision to a group of people who don't think the way you do. What is important to them and their career goals? How might your idea help them advance? It is absolutely an intelligent game.*
>
> – Yasaman Soudagar, PhD in Physics, co-founder and CEO of Neurescence[8]

Another aspect of persuasion that often causes challenges for scientists is the fact that one must often present ambitious plans to either management, an investor, or a customer. To a scientist who has spent their entire career in academic research, where it is important to be very clear about the boundaries of certainty in one's work, this can be a challenging activity. Some even feel that they are creating fiction or even being dishonest if they sound too certain about things that are only projections or plans.

In academic research, a scientist typically communicates work after it is completed. This means they are usually saying, "This is what I believe to be true. This is what I'm certain about and will stand behind." The scientist in academic research communicates certainty.

In the private sector, it is very common to communicate what the company or the team is planning to do. Investors want to see business projection before they will invest their money in the company. Managers want to see market projections and product development timelines before they will authorize funding for a new product. Customers want to see performance specifications before they will sign a purchase contract, and if they are considering a new product or technology, they may want to see those performance specifications before the product has been completed and fully tested.

Each of these communication activities requires the team to communicate results that have not yet happened. To a scientist who is new to the private sector, this can feel like they are being asked to fabricate stories. Many struggle to understand how to maintain the scientific integrity that is so central to the work they have based their careers on. But if a company is to remain competitive in their business, it is essential that they communicate their plans well before they are achieved.

In this environment, integrity is about communicating what one honestly intends to do, rather than only what one has honestly already achieved. A startup CEO presents prospective investors with a realistic, although often optimistic, plan of what their company hopes to achieve in order to secure investment capital. A development manager presents a realistic development plan to upper management in order to secure approval for the program, knowing that unforeseen things will happen that will require modifications to the plan. A company sells the new product they hope to have – the product they honestly plan to have – to a prospective customer, in order to get that critical first purchase order that will fund the final development and testing that will verify the company can actually provide what it has committed to.

Author's note: Maintain your integrity

The guidance contained in this chapter can sound a bit confusing to scientists at times. Most of us scientists began our careers in a search of truth…things that can be verified and trusted…things that can be counted on. This often comes with a high degree of integrity. Scientists want to determine what is true and state it clearly. Sometimes all of this talk about persuasion and making decisions with no right answer can sound like integrity is lost. But this does not need to be the case. I suggest you maintain that integrity!

The private sector is a game, and like every game, you will find many different styles of players. Some will be very creative and give you great ideas that you can adapt for your own playbook. As with any game, there are rules that you need to follow. Most players will be honest and play by the rules. But, as with any game, there is uncertainty. Winning requires creativity to navigate that uncertainty. You will meet players who bend the rules. You will meet players who will not play fair. You will meet players who lie and cheat in order to win.

I suggest you do not do this. Stay genuine. You're a scientist, and you begin your career believing in the value of pursuing truth. You know the value of integrity. You know how to be persistent, how to question your assumptions, and how to test your hypothesis. There will be plenty of uncertainty, and there will be times—many times—that you will not know the right answer. But do not lose your belief in the value of the truth. Don't allow people who bend the rules too much to influence you and cause you to lose your integrity. Don't let the desire to become an expert at the game cause you to lose your scientist values. These values will benefit you in the long run.

Emotional intelligence and persuasion

In the section of making decisions, we talked about emotional labor as one of three types of labor. You might consider the 'strengths' required to perform each type of labor:

Physical labor – physical strength
Mental labor – mental intelligence
Emotional labor – emotional intelligence

The term 'emotional intelligence' gained popularity after the book of the same name was published by Daniel Goleman in 1995.[25] In short, emotional intelligence is the capability of an individual to recognize and identify their own emotions and the emotions of others, and to use this information to guide their thinking and behavior.[26]

My purpose here is simply to point out that in addition to being valuable for making difficult decisions, emotional intelligence is an important aspect of influence and persuasion in the private sector as well. It's not so much that persuasion is emotionally challenging, the way making a hard decision is. It's more about the ability to recognize and understand emotions in others so that you can identify with them and see their perspective. The emotions that a person experiences dramatically shape how they see the world and therefore how they interact with it. Emotions also have a significant impact on how people make decisions.

Unfortunately, emotional intelligence is not often discussed by scientists in the context of making a significant impact with their work. But scientists who work in the private sector would do well to consider it:

> *One of my first bosses in industry once told me he would hire someone with a high EQ[21] over someone with a high IQ any day. I remember thinking that was one of the stupidest things I'd ever heard. But now I realize that you're not going to go far if you can't communicate your ideas well, and you can't communicate well with others if you can't relate to them.*
>
> – Brit Berry–Pusey, PhD in Biomedical Physics, co-founder and COO of Avenda Health[20]

Interview excerpt: Yasaman Soudagar on influence, integrity, and playing the game[8]

Yasaman has a PhD in physics from the École Polytechnique de Montréal and is the co-founder and CEO of Neurescence in Toronto, Ontario. Her full bio can be found in the Interviewee Bio section.

Dave: What are some of the challenges for scientists moving into the private sector?

Yasaman: When you get out of academia, human relations become so important. It's absolutely number one, and your technical skill set becomes number two. And human relations is something that many scientists, especially those coming from physics, need quite a bit of practice with.

Dave: That makes me think of the group video calls that I do so frequently. When the group is physics-based, no one has their video camera on. It's as though the understanding is that we discuss data so why would you need to see my face?

You came from a physics background, but now you are running a company. How have you been able to adapt and develop human relations skills?

Yasaman: That's funny, because during grad school my friends called me the queen of networking. Now I look back and think, 'Oh my god, if I was the queen of networking, everyone else must have been terrible!' [Laughs] I have learned a lot through some big failures. If I'd understood the value of human relationships sooner and had been better at working with people, things would have gone forward much easier and faster.

It took me about two years to realize the right way to apply for funding from organizations where applications are not peer reviewed. You never just send an application in by email. Most of the people judging the applications don't know the scientific details of your work, so they don't know how to judge it based on technical merit. What you do is meet with the decision maker, have coffee couple of times, and tell them what you're working on. You tell them what you are planning to propose and ask what they think about it. They will tell you if what you suggest is in the scope or not, and you listen carefully because they're telling you exactly what you should propose. Then you send the application in and you either cc that person or forward the email to them and let them know that you just submitted it. If you don't do that, it's probably not going to go through. If you've been in contact with them and kept them up to date on your work, they will have more confidence that your project will succeed, so they are much more willing to give you the money.

Dave: I was fortunate that one of my first managers in industry taught me that if you want something to happen, you need to talk to people and set it up so it will work the way you want. You can't just expect that it will work out like a formula.

Yasaman: Yes, exactly. If you are going to a meeting and you want things to go to a certain direction, you never introduce your point for the first time in the meeting. You get far better results if you talk to the key people before the meeting and bring them on board with your vision. You need to sell your vision in advance so that by the time you get to the meeting, the decision is already made.

We are always selling something, and the sooner one accepts this fact, the better. Selling your ideas is a valuable skill to have in academia as well, but it's done differently. Outside of academia, if you want to sell something to people it's important that they like you. If they don't like you, it doesn't matter how brilliant you or your ideas are. They're not going to listen to you. You need to make sure that the key decision makers like you and that you have a group who agrees with your vision backing you up.

Also, you can never just show up at someone's desk and start selling them your ideas and expect to be successful. Influence is best handled through informal meetings, and that is all about human relations. If you and I have the kind of relationship where you can come and ask me to go for a coffee and chat about the project, you will be much more successful.

Most people in society understand this, even in many fields of science, but in physics and math people often feel that building human relations is a waste of time.

A really good way to think of persuasion and influence is to think about it like a strategy game, where you have to sell your vision to a group of people who don't think the way you do. What is important to them and their career goals? How might your idea help them advance? It is absolutely an intelligent game.

Dave: Of course, science research is quite different. When we submit a paper for peer review, we don't do any human relations work or any of the advance preparation you describe. We simply send the paper in and expect that the quality of the results will do the 'selling.' Outside of academic research it doesn't work that way.

Yasaman: It really doesn't, and I can't believe how long it took me to realize that.

Dave: How did you finally come to realize it?

Yasaman: It was really trial and error. I'm sure my mentors were trying to tell me this, but it was so foreign to me that I just didn't get it. The idea that I need to have coffee with someone so they will approve my excellent application was just completely ridiculous to me.

Dave: What have you found the hardest about making the transition into the private sector?

Yasaman: The thing that I struggled with the most is integrity. In the world of business, the concept of integrity is much less clear than in academia, but it is just as crucial. There are things that work much differently than in academia, and you have to be very careful.

For example, the way you prove something is good business is with numbers, so you need to understand finance and you need to learn how to forecast. I've learned to estimate how many units we are going to sell each year and how much money we're going to make from those sales, but initially it didn't make any sense at all. Every single businessperson I talked to would tell me they knew that whatever forecast I gave them was not actually going to happen, so I thought, 'Why are we even doing this? Are we lying on purpose? As a scientist, what's happening to my integrity?' It was like being asked to fantasize about what is going to happen. As a scientist you talk about your work when it is done, not before you even start.

It took me quite a while to get there, but I realized the exercise of financial forecasting is actually simulating a company in the most accurate way possible, and its degree of accuracy depends on how experienced you are in this business. You need to learn about your target market, who will buy your product, how you will you reach those people, how will you make the product, and how will you ship it. You also need to figure out what expertise you need to hire for which roles and how many of these individuals you need. Once I got to that point, I realized that forecasting combines my love of predicting the future with formulas with statistics, that wonderful tool for managing uncertainty, to come up with a scenario for the company that we think is highly probable given our current knowledge. It is actually very far from lying! It is a very complicated activity that is meant to tell management and investors if the whole venture is worth the effort. Given this simulation of the company that is based on our best guess, will we get a good return on investment (ROI)? This revelation made me realize how mathematically and intellectually beautiful the whole financial forecast is! And of course, how necessary it is to convey the ROI to investors and other stakeholders.

When I first started my company, a couple of my friends were trying to teach me to think about the potential of my technology and how to sell it, and I just couldn't get behind it. My attitude was, 'We are building a brain imaging device. I can't go in front of these investors and tell them we are going to completely understand the brain and solve all brain problems in 10 years. I know that's what they want to hear, but I know what we can and cannot do as humans! You are telling me to go out there and lie, and I just cannot do that.'

Their position was, 'If you tell a story about the *potential* of your device, that's not lying.' They asked me, 'Are you telling me that there is *no chance* that your technology will ever help us find a cure for Alzheimer's?' I replied, 'Of course there is a chance that will happen. Otherwise, what am I doing here?' And they said, 'Then you are not lying!' This was so hard for me, because in science we need to be 100% certain before we claim something.

Then a friend who works in TV gave me some advice that really helped. Her job at the time was to translate science documentaries like those on the Discovery Channel into French for the Canadian TV channel in Quebec, and so she was able to see more of my perspective. She told me, 'You keep saying that you're a scientist and you can't lie. I was just working on a documentary on stem cell research. It's a scientific documentary, but instead of just telling you about what is happening in the petri dish, they keep telling you about the potential... about what they *hope* it is going to do for humans. That's what you need to do. You are selling hope!'

I'm a much better storyteller now. I share personal things that I never would have before, because I've learned that to persuade people, you need to connect with their guts, not their brains. That was the mental shift that I needed.

Dave: I struggled with that as well, because you are always selling something you haven't done yet. I came to think of it as selling the product that I planned to have... the product that I hope to have. As long as I really do plan to make that happen, I am being honest.

Yasaman: It is a challenge, but I can tell you one thing for sure. In business there is a line, and people who cross it will go down. They might be successful for a year or two but eventually it's going to come out.

Dave: I used to work with a sales guy who sold products based on claims that were just not true. He actually used to say, 'I know what a liar looks like, because I see one every day when I'm shaving.' He made an initial sale to many customers, but it all caught up with him when nobody wanted to buy a second.

Yasaman: Exactly. And in addition to your honest intentions, you also need to be sure to communicate in a way that doesn't cause any misunderstandings. It can be difficult for those of us in science, because we are used to going to the board and writing math formulas to communicate what we mean. In these interactions we are all talking the same language, so it's easy to be sure everyone understands. People in business come from all sorts of backgrounds so a statement that means something to me may mean something completely different to another person. If a miscommunication is interpreted as dishonesty, then it can be very hard to repair that relationship. But if you are being genuine, people generally see that. If you make a genuine mistake or if the miscommunication was due to a language difference, those aren't big problems. This understanding comes with experience, from talking to people, making mistakes, and realizing what caused the misunderstanding.

Successful industry habits:

1. Help the company make money
2. Figure out what matters and what doesn't
3. Be effective, not smart
4. Decide quickly with limited data
5. Persuade others to follow you

TurningScience

Chapter 5
The R&D Mindsets

The versatile scientists the world needs

In the previous chapters, I've outlined the differences between the academic research environment and the industry environment. While emphasizing the distinctions are critical for helping scientists transition from the academic environment into the private sector, I want to avoid giving the impression that a successful scientist must focus on only one set of habits or the other.

I also want to avoid an 'us versus them' mentality, which is ultimately counterproductive and yet far too common in both academia and industry. Most PhD scientists are aware by the time they complete their postgraduate studies that many science professors project a negative attitude about working in industry, and there are also many negative stereotypes about the academic research world that permeate industry. These attitudes are a natural result of our human tribal tendencies, but they are not useful.

A far more valuable perspective acknowledges the different goals of each environment; recognizes that the people who work in each may have different strengths, interests, and habits; and focuses on the value of healthy cooperation and even collaboration. Communities where the local university and local industry have recognized this value and have worked to encourage cooperation have shown significant benefit to both.[1]

Academia and industry are both critical for the advancement of humanity: industry relies on the new science that comes from academia, and academia benefits from the tools created in industry. Neither can achieve its full potential without the other—and humanity achieves the best outcome when both are equally respected, and the people in each field work together to create new knowledge and new solutions. The most versatile and influential scientist understands both environments and can function well in either one.

Research and/or Development?

The term 'research & development,' often shortened to R&D, is used in the private sector to describe two important activities in the creation of new technology solutions. These two terms form an excellent basis for describing the differences between academia and industry environments, but also the basis for identifying where the two environments can benefit from similar perspectives.

The two elements of R&D, 'research' and 'development,' are quite different activities, with the former focused on the creation of new knowledge and the latter focused on the creation of solutions. The majority of the work that might be termed 'R&D' in the private sector is typically focused far more on development than research. Conversely, most academic groups are focused almost exclusively on research, although there are many great examples of commercially focused development work that resulted from private-sector collaborations.

It is valuable to consider that there is a different mindset for each environment. One is the research mindset, typical of academic research, and the other is the development mindset, which is far more prevalent in industry. However, I prefer to think of each mindset as associated with the activity that each describes—research or development—than with either academia or industry. This is because both academia and industry benefit from both mindsets, although in different measures.

The most valuable scientist understands and embraces both mindsets and learns to recognize which is needed in any particular project or activity. This will make them more versatile, more influential, and open them up to a much wider array of career paths and opportunities than simply embracing either the 'academic' or 'industry' mindset and working habits.

It is important that we acknowledge the valuable synergy that can be achieved through utilizing both mindsets in either environment. After all, why can't a scientist contribute to both the creation of new knowledge and to the creation of new solutions in a single career?

> *One of the biggest challenges (of my transition to industry) was understanding the difference between a research project and product development. The term Research & Development used in industry suggests that both of these are done in tight conjunction. However, that does not mean the activities are similar. They involve very different processes and mindsets that often conflict with one another. The cool part is, once a scientist knows their way around research AND development, they turn into absolute power players.*
>
> – Oliver Wueseke, PhD in Molecular Biology, founder and CEO of Impulse Science [2]

The R&D Mindsets

In short, the research mindset is focused on generating new knowledge and the development mindset is focused on quick progress toward a solution.

How might we clearly describe the two in more detail so that it is easy to see which is needed in any given situation?

Consider the following:

		Research Mindset	Development Mindset
Characteristics	Primary goal:	New knowledge	New solutions
	What is pursued:	Understanding	Results
	What is rewarded:	Certainty	Speed
	How progress is made:	Proof	Persuasion
Behaviors and Habits	Decide based on:	Proof	Probability
	Focus on:	Completeness	Efficiency
	Response to uncertainty:	More data and more analysis. Sounds like: 'I'm not certain, so let's keep looking.'	Decide quickly and test your choice. Sounds like: 'I'm not certain, so let's try something.'
	Integrity means:	Accuracy: What we honestly think is true	Plans: What we honestly hope to achieve

Playbook Sketch 6 Characteristics, behaviors, and habits for the research and development mindsets.

When considering an activity in either environment—academia or industry—consider the primary goals of the activity. If the primary goal is learning, the research mindset is likely the best approach. This is useful for activities where a complete understanding is desired and certainty takes precedence over speed. In these activities it is generally acceptable to take as much time as required to achieve a high level of confidence before proceeding to the next step.

If the activity is primarily developing a new solution to a problem or a device that serves a specified purpose, then the development mindset is likely the most valuable. The primary objective will be some predefined result, and there is usually value in moving quickly because the thing that is created is needed for some other purpose. Sufficient performance to achieve the desired results can generally be achieved without absolute certainty, so it is usually valuable to make decisions based on limited information and test as you move forward.

Why embrace both mindsets?

One may ask, "Why is it important for a scientist to understand both environments and incorporate both the research mindset and the development mindset into their work? Can't one pick one area to build their career and focus on the behaviors that bring success in that environment? Isn't that more efficient than learning and trying to use both?"

It is certainly common for a scientist to work exclusively in either academia or industry for their entire career. Many people find that they enjoy working in one environment significantly more than the other, and often find that their personal strengths make them better suited for one or the other as well. I have certainly found this to be true for myself. The time I spent in academic research during my PhD work was very enjoyable for me, but five years was enough time to spend in a pure research environment. I found that I preferred the goals and attitudes I found in the private sector, and my strengths are much better suited for industry than academic research.

But it's not as simple as saying that if you work in academic research you only need one mindset, and if you work in industry you only need the other. One mindset is much more important than the other in each environment, but I have found that even if one chooses to work exclusively in either academia or industry, there are many scenarios where they will find value in understanding and incorporating behaviors from both environments into their work. It is also not uncommon for a scientist to consider a career transition later in their career, and it will be beneficial for them to have cultivated an understanding of both environments throughout their career.

> *Despite the fact that I'm now working in a business environment, I don't feel that my science career has ever ended. As scientists we are trained to look at a problem, hypothesize about the fundamental issue, and draw conclusions. This is baked into our existence. Yet business is far from an exact science. Try dragging and dropping a little scientific intelligence, creativity and intuition into business and I'm sure you'll be pleased with the results.*
>
> – Scott Sternberg, MS in Physics, Executive Director of the Boulder Economic Council [3]

Playbook Sketch 7 outlines a few such scenarios where the use of both mindsets may be valuable.

The Versatile Scientist

The scientific method is by far the most powerful approach we humans have to achieve consensus regarding the universe and how it operates. The scientist the world needs most is one who can bring the power of the scientific method to the world outside academic research, and this is best done when they can

Scenarios where both mindsets are needed

Academia

- Research group builds a collaboration with a private-sector company
- Professor trains graduate students in industry habits to improve their private-sector employability
- University researcher creates a startup to commercialize new discovery
- Academic research group faces tight deadlines to demonstrate results due to funding restrictions or competition with other groups

Industry

- A company invests in researching new science to support a new product direction
- Technology transfer from a university research group to a company
- Industry manager engages a professor as a subject matter consultant
- Senior scientist mentors new Ph.D. scientists as they transition into product development

Career Transitions

- Technology transfer from a university research group to a company
- Industry manager engages a professor as a subject matter consultant
- Senior scientist mentors new Ph.D. scientists as they transition into product development

Playbook Sketch 7 Scenarios where both research mindset and development mindset are valuable.

operate in both mindsets. They have the intelligence and the skills to use the scientific method to identify the right answer when possible—and they have the perspective and the emotional intelligence to make decisions when there is no right answer, and persuade people to follow their ideas and recommendations. The scientist who understands and embraces both mindsets is the scientist who can operate well outside of academic research, working alongside non-scientists in all areas of life. They are the scientists who can build bridges between scientific research and the rest of the world.

Bridges are connectors, and connectors are very important. Every project has several different subsystems. The subsystems can have the best design in the world, but if we don't have a good connector between them, and if they are not communicating well with each other, the overall project cannot be successful. Performance of the entire system depends on how good the connectors between the subsystems are.

– Sona Hosseini, PhD in Engineering; Applied Science, Research, and Instrument Scientist at the Jet Propulsion Lab[4]

The development mindset in academic research

The best way to foster both R&D mindsets in scientists is to teach and encourage both mindsets during their PhD training. Practicing both mindsets while in university will help PhD scientists become familiar with both and learn to determine when each is appropriate in whatever career path they choose. And for those who chose private-sector careers, it will help them be more valuable as members of an industry R&D team.

While the research mindset is certainly the most applicable in academia, there are many opportunities to foster the development mindset in an academic research lab. But this is not simply a suggestion for training purposes. Bringing the development mindset into the academic research environment has significant benefits for the team and for the results they achieve as well.

I'm often asked by professors and university administrators how they might best help their PhD students and postdocs transition into industry when they leave academia. The R&D mindset perspective offers a great way to think about training new scientists for their careers outside of academia. I suggest that they develop an awareness of the R&D mindsets in their labs by encouraging these three habits.

Habit #1: Move fast and decide quickly

The most valuable habit to encourage is one of moving fast and making decisions quickly. When a research group is instilled with a sense of urgency, it will naturally drive them to adopt the industry playbook habit introduced in Chapter 4 of 'figuring out what matters and what won't matter' in pursuing their research goals. Stopping to analyze a project plan—and determining what is critical for achieving the goal and what is not—is always a good idea for promoting efficiency, whether the goal is commercial product development or fundamental science research. Working for efficiency in academic research need not result in the loss of accuracy of the results or completeness of the project. It will counter the tendency of many academic researchers to work slowly and deliberately through the project, spending more time than necessary on less-critical aspects of the project.

Instilling a sense of urgency also naturally drives the group to 'decide quickly with limited data' on non-critical decisions, which is the fourth habit from the industry playbook presented in Chapter 4. While some decisions in academic research may have far-reaching consequences and deserve careful analysis and consideration, most groups find that this approach need not be applied to all decisions. There are many scenarios where the development-mindset principle of 'decide quickly and test your choice' will save valuable time. Building experimental hardware may be one such scenario. In industry, the practice of 'prototype early and prototype often' is the time-tested way to quickly arrive at the best choice for a new product. In academic research, this practice will reduce procrastination by encouraging scientists who tend to look for certainty before proceeding to the next step to make a decision and move forward, knowing they can adapt in the future if needed. The time that is saved from many decisions made quickly makes up for the possibility of lost time to occasionally correct a 'prototype trial' that didn't work out so well.

My advisor used to say, 'Any job worth doing well is worth doing fast.'

– Jason Ensher, PhD in Physics, Executive Vice President and Chief Technology Officer at Insight Photonic Solutions

---

Habit #2: Encourage teamwork

One of the most valuable things that a PhD signifies is the ability to do independent research. This trait results from the fact that most PhD projects are completed entirely by the candidate themselves, so that through the process they become an expert in all aspects of the project. This has value and should remain the basis of the PhD dissertation project, but there is great value in encouraging teamwork.

In discussing teamwork, I want to differentiate between two different scenarios: (1) An environment where the PhD candidate works alongside other researchers who may share their challenges and accomplishments with each other but where the success of the candidate's project remains primarily a function of their own efforts, and (2) True teamwork, where the PhD candidate must rely on the input and assistance of others to accomplish their own goals. Scenario 1 is common in many research lab environments (it was certainly my own experience), and it encourages independent work habits, but it does little to help prepare scientists for the team environment of the private sector.

As many successful academic researchers already know, academic research groups can benefit significantly from an environment where the researchers depend on each other for some critical aspects of their work. In addition to developing collaborative work habits that will benefit those who take their careers into the private sector, the intra-group communication will be improved, problem solving will become a collaborative effort, and the researchers feel more invested in the success of the whole group than only their own projects.

One way to foster true teamwork in academic research is for the group leader to assign a specialization to each member of the team. In this model, each researcher becomes an expert on some process or capability that is used by the entire team, encouraging close collaboration in order for team members to complete their own projects. This model benefits the group in that it allows each person to become the expert in some capability and work to optimize it to a degree that each researcher using it for only their individual projects could not. Each researcher is then required to train the rest of the team on their expertise area to a sufficient level of detail for completing their dissertation and passing their thesis defense.

The basis of true teamwork is the realization of one's own limitations and the acceptance that help from others is the fastest route to progress. A PhD advisor or principal investigator (PI) can set a very important example by admitting when they don't have the answers and encouraging a team member to look outside the group for help.

Habit #3: Think of your research group like a company

This idea came from my own PhD advisor during a conversation we had right after a lecture I gave at my alma mater on private-sector careers. In response to the five 'habits successful scientists in industry learn quickly' that I had presented, my advisor suggested that I encourage PhD students to think of the research groups like a company, an organization with a plan for growth and goals and objectives that reach beyond those of the individual members. She proposed that PhD students who focus on the needs of the whole group, rather than only their own project, can shed the stereotypical PhD habits and foster new habits closer to those critical for success in industry. A PhD student who is invested in the goals and objectives of the research group is more likely to focus on teamwork and achieving fast results that help the group. The student is more likely to focus on what results bring future value for the group. Should that student decide to pursue a career in industry, the habit of focusing on what achieves value for the group will easily transform into the habit of focusing on what helps their company make money.

I thought this was a great idea and began to look for examples of this from other science research groups. I recalled that two of the interviewees from my first book, Jason Ensher and Chris Myatt, had described how they emerged from university with several development-mindset habits, despite coming from groups that worked on fundamental science research. Both of these scientists have since gone on to very successful careers as senior executives and founders of successful technology companies.

Jason and Chris worked in two different groups that were collaborating in pursuit of the first demonstration of Bose–Einstein condensation, for which their advisors ultimately received the 2001 Nobel Prize in physics.[5] I found it so interesting that scientists who were trained in academic research groups that had achieved what most consider to be the pinnacle of academic research success had developed habits that were so 'industry-like.' But while the five 'habits that successful industry scientists learn quickly' are important for success in the private sector, they can be quite valuable in academic research as well. The fact that they were competing against other groups to be the first to achieve an important goal drove them to be efficient in their efforts and focus on working as quickly as possible. These are essentially the same competitive pressures that make the industry playbook habits so important in the private sector.

> *It's also very important to focus on the problem at hand and not waste time on non-critical details. My dad was a very practical thinker and I picked that up from him. This attitude worked well in graduate school with an advisor like Carl Weiman. His opinion was if you are getting more than a B in your classes you are not spending enough time in the lab. Every physics student wants to talk about electrodynamics and field theory, and other cool things but his attitude was "Get back in the lab and get the experiment working."*
> – Chris Myatt, PhD in Physics, founder and CEO of LightDeck Diagnostics[6]

The stories of Jason and Chris demonstrate that the development mindset is valuable for achieving ambitious goals in academia as well as helping a scientist transition effectively into the private sector.

Interview excerpt: Oliver Wueseke on his advisor's valuable example[2]

Oliver has a PhD in molecular biology and is the founder of Impulse Science, where he mentors startups in research strategy and product development. His full bio can be found in the Interviewee Bio section.

Dave: Trying to be the expert in everything is not expected or even encouraged in the private sector, but I was well into my first job before I realized that my habit of trying to be the expert in everything just didn't work anymore.

Olli: There's one pivotal moment during my PhD that I remember helped me to make that shift. I had the pleasure of working with some exceptional scientists that were not only good scientists but also empathic leaders. There was this time when I was struggling with a biophysics problem in the lab. I was stuck in the PhD 'Valley of Tears' and didn't know what to do. I went to my supervisor for help and admitted that I had no idea what was going on. He said, 'I think you're doing great. I do not have the answer that you are looking for, but let me put you in touch with somebody who can help you out.' That response, coming from somebody that I looked up to, was very powerful. First, he was empathic about my situation and gave me comfort. Then he acknowledged that he wasn't perfect and all-knowing. Lastly, he pointed me in a direction that allowed me to solve my problem. His authenticity in this situation had a massive impact on me. It allowed me to accept my imperfection and showed me the value of actively seeking out help from others. Rather than trying to know everything, I now profoundly invest in personal relationships with people that have more experiences than I do!

Interview excerpt: Jason Ensher on the development mindset in academia[7]

Jason is a PhD physicist who, like Chris Myatt, earned his PhD in the labs of Nobel-Prize-winning physicists Eric Cornell and Carl Weiman. When I interviewed Jason back in 2010, I was so impressed by some of the industry-ready habits that he possessed when he transitioned into industry. Jason's full bio can be found in the Interviewee Bio section at the end of this book.

Dave: What are some of the things you learned in school that helped you make this transition?

Jason: Above all, I learned the value of working quickly. My advisor used to say, "Any job worth doing well is worth doing fast." I have had people debate that with me, but if you get the result you want, any more time spent on it is wasted. This also pushes you to explore the parameter space of failure because no matter how smart you are, you are more likely to fail than you are to succeed. Work quickly and figure out what doesn't work so you can find what does.

I came to industry with a very questioning attitude, more so than the other engineers who had been working there for a long time. I questioned what the product had to do and what the development effort had to achieve. I asked the marketing people about what the customers wanted. If one of the engineers indicated that we needed to build a circuit to do something, I would ask why. I wanted to know that we were being efficient and working as quickly as we could.

Second, being able to demonstrate working hardware is so important, even if it is not working perfectly. It is often necessary to demonstrate some basic level of functionality in order to get management to fund the next stage of a project or to demonstrate capability to a customer. In graduate school, I learned how to interpret 'bad' data, i.e., explain the effect of a certain piece of test hardware that wasn't working properly that day. This is exactly the kind of approach to take when demonstrating a prototype product. Having the ability to explain to your boss or a customer why it is not working to specification and what you can do to fix it is so much better than saying, "I don't have something that works for you now."

What has always gotten me the most traction in industry is being able to show data—hard facts—not just analysis. Theoretical analysis can be very important, but you can spend a lot of time on it. If you have the skill to perform an experiment in one day that demonstrates what you are talking about, you can save weeks of time. I have found that evidence almost always settles a dispute, even if the disagreement is political. Data-driven conclusions are very important.

Finally, I think it is very important to be able to talk intelligently about the work you are doing. My father was a technical salesman, and he emphasized to me that you need to be able to talk about what you are doing or you won't convince people how good your product is. I had great coaching on this in graduate school. The students would give talks in front of a big group, and Eric and Carl would critique us, often on the language we used to describe our work. As a scientist and graduate student, you tend to become comfortable talking about your ideas and having them challenged, and this skill is very useful in industry. I've seen many scientists who were successful in business development or marketing because they had these skills.

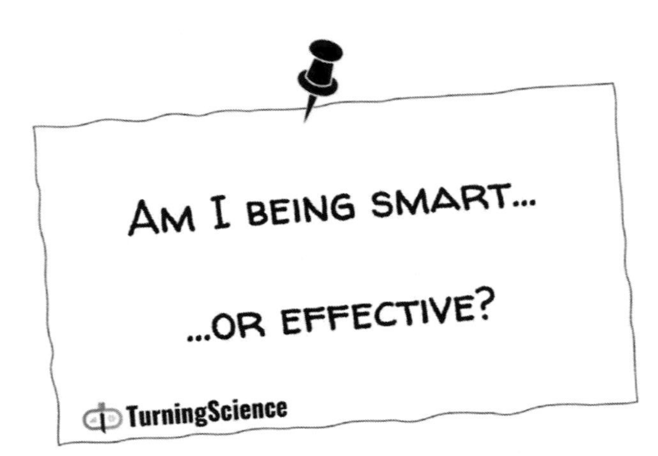
Am I being smart...
...or effective?
TurningScience

Chapter 6
Startups – The Ultimate Game!

Want to make your own dent in the universe?

Have you ever thought about becoming an entrepreneur? Many scientists choose to pursue the challenge and excitement of starting their own company at some point in their career. Many more scientists choose to join an existing startup to help commercialize some new cutting-edge technology.

But is a scientist a good fit for the startup world? Can a scientist really be successful as an entrepreneur? Many people consider the skill sets needed to be a successful scientist to be very different that those of an entrepreneur and assume that a scientist would not be good at starting their own company. After all, scientists figure out how the universe works, and they like to study hard problems for years, running experiment after experiment, until they find the right answers. That's very different than building a business. Companies do not have years to look around for the right answer. They need to move quickly, make decisions, and develop solutions for their customers, or they will not survive. So, a scientist is probably not well suited for launching their own company, right?

That's what I used to think as well. But as I reviewed the interviews of successful scientists in industry that I conducted to select the ones I would put in my first book, *Turning Science into Things People Need,* I found that half of the most engaging interviews were with scientists who had started their own companies. Five of the ten industry scientist interviews that I put in my first book were with entrepreneurs, all of whom had survived for more than ten years after founding their companies. And all of these scientist entrepreneurs credited their science background as an important part of their success.

Those interviews, along with all of the work I've done with entrepreneurs in the last ten years since that book was published, have convinced me that scientists can actually be great entrepreneurs. Of all of the stories I've collected from successful scientists in industry, the 'scientists-turned-entrepreneurs' have given me some of the most valuable advice for playing this game of 'turning science into things people need.' A scientist who becomes an employee in a larger company might manage to survive despite not being a very good performer, particularly if they have a forgiving boss and a technical expertise that is valuable to the company. But a scientist who launches their own company must learn the game quickly and perform at a very high level or they will fail. Their advice has benefited me many times, and it forms the basis of much of what I teach in my industry-career workshops.

This chapter doesn't try to tell you how to be a successful entrepreneur, as a wealth of information to help you with that is already available in other publications.[1] My focus here is use the stories I've collected to outline the strengths that scientists typically have that can make them good entrepreneurs, to highlight some of the common challenges that are unique to a scientist who chooses to play the startup game, and to relay a few 'playbook' suggestions to help scientists avoid these challenges.

If you've ever considered 'taking the leap' to start your own venture, or even if you think that joining someone else's early-stage technology company might just be the next step in your career, this chapter will help you understand where your scientist strengths can be valuable in playing the ultimate game!

It's about being able to shape the world, and that's very cool.

– Chris Myatt, PhD in Physics, founder and CEO of LightDeck Diagnostics[2]

Why Become an Entrepreneur?

The stories of scientists-turned-entrepreneurs begin with the important question of how and why they arrived at the decision to start their own companies. While the specific circumstances vary in each story, there are some common themes behind why scientists choose to launch their own venture.

Appetite for challenge and adventure

The stereotype of the scientist that exists outside of the scientific community is a person who approaches their life and work in a careful and deliberate manner, and is hesitant to take risks. While in reality this may describe some scientists, it certainly does not describe all of them. Here are two scientist entrepreneurs who clearly do not fit that stereotype.

Tom Baur, MS in Astrophysics

In 1979, Tom Baur was an astrophysicist working at the National Center for Atmospheric Research (NCAR) in Boulder, Colorado, where he spent time doing observational astronomy and developing instrumentation for solar physics experiments. Growing up in farming had cultivated a strong interest in practical applications, and Tom found the instrument development aspect of the job more fulfilling than the pure research. Tom also found that a career in research didn't provide the challenge and the independence that he craved. Tom's work in instrument development convinced him that there was a market for optics that control the polarization of light, so after 13 years at NCAR, he left and started Meadowlark Optics in a spare bedroom of his home. As Tom describes it:[3]

> *Starting my own company was all about my need for risk and challenge. I could not satisfy that need in a structured research environment where credentials seemed to be more important than achievement. I also felt that the link between performance and funding was not clear. What you accomplished seemed to matter less than the current political environment. I need more of a connection between risk and reward. I also need to feel useful, and that was lacking for me in pure research. The link between mapping magnetic fields in solar active regions and helping mankind was too loose for me.*

Today, Tom is still the chair of the company he started 42 years ago, and the risks he took in leaving the security of his job of have resulted in many great rewards:

> *It's been pretty scary at times, but very satisfying... I began Meadowlark as a bootstrap operation, and we still have not taken any extra money. We developed a line of products that are useful to people, and that's important to me. I am also proud that I provided employment for lots of people and helped them grow in their careers.*

In addition to providing solutions for many customers and employment for many employees, Tom has made a significant contribution to the education of future scientists. In June of 2020, Tom and his wife, Jeanne (together with SPIE and the University of Colorado, Boulder), established the first Endowed Chair at the Joint Institute of Laboratory Astrophysics (JILA).[4]

Peter Fiske, PhD in Geological and Environmental Sciences

As the son of a geoscientist, Peter Fiske grew up with a very strong science influence. From an early age Peter decided he would follow in his father's footsteps. But he was more of an extrovert than the typical scientist and not particularly good at math, so he realized early on that being a research scientist was likely not the best fit for his strengths and interests. Peter completed a PhD in geological and environmental sciences from Stanford University and then accepted a postdoc position at Lawrence Livermore Labs in Livermore, California, all the while looking for alternatives to the traditional academic career path:[5]

By the end of my postdoc, I decided that I wanted to pursue a career in science policy, so I applied for and got a White House fellowship. This is a program that recruits early- to mid-career people from all walks of life to work for a year as a special assistant to the President, Vice President, or a cabinet secretary. I headed off to (Washington) D.C. to start my new career, but after two months I realized I had made a terrible mistake. The pace of progress in the federal government was just too slow. I soon realized I wanted something a lot more entrepreneurial, so after a great year as a White House fellow, I went back and got a staff position at Lawrence Livermore Labs.

While Peter's experience as a science assistant for the U.S. government did not provide a viable career path, it opened his eyes to new and different opportunities. Peter continued searching for something that fit better while working at Lawrence Livermore:

I stayed at Livermore for four years, but during that time I was also poking around in entrepreneurship. I enrolled in the evening MBA program at the UC Berkley School of Business and in my second year started fishing around for ideas for Berkley's business plan competition. I had this great idea that I would be an agent to link technologies at Livermore Labs to the Business School at UC Berkley. I went around the lab, gave a series of lunchtime talks about entrepreneurship and the business plan competition, and asked anyone who was interested to give me a call. I got around eight calls from people. Four were pretty good, and one stood out. The one that stood out was an optics manufacturing technology that had essentially been abandoned by the lab because it was too high risk.

I worked with the inventor, and we decided to found RAPT Industries. Then I wrote the business plan and entered it in the business plan competition. I still remember vividly the night they announced the winners. The first thought in my mind when they announced that we had won first place was, 'Oh, @#%&! I am going to have to quit my job!'

So, I quit my job and started working full-time to get RAPT up and running. We took in a small round of initial investments and then leveraged that with a lot of government R&D funds. That is what I did for the next six years.

Through some creativity and perseverance, Peter created an opportunity that fit his strengths and interests much better than the traditional academic science career. Becoming the founder and CEO of a technical startup proved to be an exciting alternative that satisfied his love of science and leveraged his strengths as a leader and communicator.

Making a difference in the world

Many science PhD students face the realization at some point during their postgraduate career that the problems they are working on are not going to lead to the kind of world-changing discoveries that they read about in their coursework. Most of us end up working on a very detailed aspect of a niche project that will never end up in a textbook. The realization that the publications we produce as a graduate student may only be read by a handful of other scientists working in our specific sub-topic causes some of us to look for career opportunities that have the potential for greater impact.

While many scientists find that a career as a company employee brings them close enough to the impactful practical applications that they longed for, some desire the challenge and risk/reward connection that Tom Baur described above, and so they set out to put a 'dent in the universe' from something of their own creation.

Marinna Madrid, PhD in Physics

While Marinna Madrid was a graduate student in the physics department at Harvard University, she and her lab partner, Nabiha Saklayen, developed a novel method of tissue engineering using lasers:[6]

> *We co-invented several laser-based techniques for intracellular delivery, figuring out how to use lasers to poke tiny holes in cell membranes and deliver cargo into the cells. For example, one experiment we did was working with induced pluripotent stem cells that were expressing green fluorescent protein. We figured out how to deliver CRISPR-Cas9 into those cells to knock out the gene that was responsible for expressing green fluorescent protein. So basically, before laser delivery, the cells were glowing green, and after laser delivery, the cells were not glowing green.*

The two scientists suspected that their research could be used to solve problems people were currently facing, and that prospect interested them:

> *In the last year of my PhD, (Nabiha and I) decided to create this startup with our third co-founder, Matthias Wagner. He's a very experienced optical engineer and also a serial entrepreneur.*
>
> *A critical moment in our decision to start the company was the startup competition that SPIE holds every February at the Photonics West Conference. We had decided on a whim to apply to the competition, but our 'startup' was really no more than a proof-of-concept demonstration in the lab at the time. We still hadn't figured out what our market or target application was yet. I remember thinking that*

we weren't going to do very well because everybody else in the competition was much older and appeared very experienced, and their startups were way farther ahead of ours.

But I gave the pitch and we ended up getting first place and some money as a prize. Once you have funding, a startup becomes a reality. That was really validating for us and encouraged us to keep pursuing it.

So Nabiha spent the entire summer after the startup competition doing customer discovery, asking biologists who found our technology interesting what they might use it for. I spent that time in the lab doing proof-of-concept experiments based on those conversations. From that, we convinced ourselves that what we were doing could be applied to interesting problems that people actually care about.

The company Marinna and Nabiha founded in 2017 and still work at today is called Cellino Biotech.[7] At the time of my interview with Marinna, she and Nabiha led a team of nine people working to use lasers, robotics, and machine learning to build a fully automated approach to cell engineering. A particularly useful application of Cellino's technology is manufacturing tissues for regenerative medicines, making it highly likely that Marinna's research will be making a real difference in people's lives in the next few years.

A lot of PhD research only seems useful to the people working on it, but winning the startup competition showed us that this technology could potentially be useful to the outside world.

– Marinna Madrid, PhD in Physics, co-founder of Cellino Biotech

---

Chris Myatt, PhD in Physics

The path that physicist Chris Myatt followed to start his company, LightDeck Diagnostics, was very different than Marinna's, but his primary motivation as an entrepreneur is very similar. Chris was already an entrepreneur at the time, having founded Precision Photonics Corporation (PPC) in 2000 with his wife, Sally Hatcher. PPC was created to develop complex optical components, and they quickly found a very robust business in the telecommunications market. However, when the dot-com bubble burst in the early 2000s, the telecom hardware business declined as a result. Chris was challenged to find another market to pursue:[2]

It started in 2003 when telecom had just gone to hell, and Milton Chang, one of our investors, said, 'You've got to go find something else to do.' We deliberately sat down and thought about what would be the coolest

thing that I could do as a measurement guy, combined with the skills I had developed within the company. Low-cost medical diagnostics seemed like the right thing.

The new medical diagnostics project proved to be the right choice, and mBio Diagnostics was spun out as its own company in 2009. In 2020, mBio Diagnostics was rebranded as LightDeck Diagnostics, and it currently develops *in vitro* diagnostic tests based on an innovative waveguide platform that enables 'lab-quality results anywhere, in minutes.'[8] In late 2020, the company received an investment of $11 million to develop a rapid test for identifying SARS-CoV-2, the virus that causes COVID-19.[9]

While mBio was conceived as a lifeboat to save his first company, Chris receives satisfaction from making a difference and solving important problems that humanity faces. This is in sharp contrast to how he feels about the science research he did as a graduate student. Chris was fortunate enough to work on the team at the University of Colorado–Boulder that first demonstrated Bose–Einstein condensation (BEC) in 1995, for which his advisor, Carl Weiman, was awarded the 2001 Nobel Prize in physics. Despite working on such a fundamental demonstration of an important physics principle, Chris was much more drawn to creating solutions than demonstrating theories:

Scientists tend to have little connection with the applications for their technology. They may know why it matters from a scientific perspective, but not necessarily why it matters to society... Early on, I was lacking this connection to the real world, as well. When we achieved BEC, the world didn't change. Most of the results were things people could sit down with a pad of paper and predict. When that finally dawned on me, I was a little bit disappointed. Around 200 academic groups have jumped on the BEC bandwagon, but so what? What will be much more satisfying to me is if I see our medical products in clinics around the world in the next 10 years. The concept of being able to make a difference in the world is extremely rewarding.

The right opportunity

In addition to having a desire to make a difference in the world and an appetite for the challenges that come with it, becoming an entrepreneur requires the right opportunity. Although most of the entrepreneurs that I've talked to seem to make a habit of creating the opportunities that they want, it is still worth noting how they accomplish this.

Brit Berry–Pusey, PhD in Biomedical Physics

Brit Berry–Pusey realized early in her life that her love of science and desire for challenge might both be satisfied by the same career path:[10]

In high school, I loved math and science, and it was really important for me to be challenged. When I was growing up there weren't that many women in science. I wanted to be different, and I also didn't want anyone to be able to figure out what I did just by looking at me, so that drove me into science as a career.

I started college when I was only 16, and someone told me that physics was the hardest subject. That's why I decided to become a physicist. I completed an undergraduate degree in physics, but I still wanted to help people, so I completed a PhD in biomedical physics at the UCLA school of medicine.

Brit also knew from an early age that she wanted to make a difference in the world. Her earliest aspirations were to be a doctor, but as she progressed into her education, she began looking for how to make the most significant impact as a scientist:

I knew I didn't want to go into academia when I started graduate school, so I chose my graduate program based on becoming a medical physicist. At the time, I imagined I would end up working with patients in a hospital developing treatment options. But in my second year in graduate school, I took a class about becoming an entrepreneur in science. The instructor, Roy Doumani, was a prolific businessman and very active in helping scientists commercialize their ideas. We really connected, and he became an important mentor to me.

I learned that if you want to have a major impact on the world, you need to take your ideas out into the marketplace. I once heard, 'If you want to help thousands of people, become a physician. If you want to help millions of people, become a scientist and make sure your ideas get out into the world.' I realized that the largest impact you can have is if you translate your ideas out into the world, and to do that, you need to be thinking about the business applications of your ideas.

This was a very new experience for me at the time, because neither of my parents were businesspeople, and I had little knowledge what business was about. My mother was a schoolteacher, and my father was a pilot, and I grew up with this concept that business was a bad thing where people go to screw other people over. When I took the entrepreneurship course, it really opened my eyes and completely changed my perspective on what I could do as a scientist and how to get ideas out in the world to help people.

Roy was a serial entrepreneur and investor. Because he and I really hit it off, he ended up hiring me to do the technical due diligence for the investments he was considering. Through that work, we decided to start a program at UCLA called the Business of Science Center. We brought students in from all across UCLA—medical students, engineering students, law students, and business students—to work together and evaluate ideas coming out the university and try to find market opportunities for commercialization. It's a really great program, and it has opened so many students' eyes to new opportunities. By the time I left UCLA, we had over a thousand members in the Center, and it's something I'm very proud of.

By taking the initiative to explore non-traditional career options, Brit was exposed to new ideas and to people with valuable experience to share. By reaching out and seeking advice from the people she met through her explorations, Brit made many solid contacts who knew more than just her name. They got to know her strengths and her ambitions. This made her more visible to the influential and resourceful people she crossed paths with, creating the opportunity to do big things like helping create the Business of Science Center. And by having the courage to take on such a challenging project that exposed her to an even wider range of influential people, she put herself in a position to impress many others with her initiative and potential.

What may seem to some as a 'lucky break' was actually the result of many courageous steps that resulted in creating that unique and valuable opportunity.

I met a lot of really great contacts during that time. In fact, I first met one of the co-founders for the company that I am running today through the Business of Science Center. He ended up contacting me 10 years later when he was starting a company based on intellectual property he had created, and he asked if I was interested in starting a company together. I jumped at the chance, and that company is the one I'm working with today. It all happened because of connections that I made through the Business of Science Center.

Yasaman Soudagar, PhD in Physics

Yasaman Soudagar had planned a career as an academic researcher, but as she was finishing her PhD, a friend told her about an opening at the company where he worked that he thought would be a great fit for her. While Yasaman did not want to change her long-term career plans, she was curious what working in industry was like. She convinced them to create a temporary industrial postdoc position for her so that she could return to academia after a few years, but found industry so exciting that she decided to stay.

A few years later, Yasaman was developing laser delivery optics for surgical applications at a company in Toronto, Canada, when she saw an opportunity:[11]

One day, I learned about in vivo imaging of neurons from a friend who worked in neuroscience. I had already seen images of neurons by a two-photon microscope at the lab of Dr. Tom Baer at Stanford and had become fascinated by it. I realized I could make an in vivo optical imaging system that was better than the existing ones on the market. So, I disclosed the idea to the company. It wasn't really aligned with the company's interests, so they agreed to assign the IP to me. I spoke to several of my mentors, including Tom, and they suggested that I start my own company to pursue it. That project became Neurescence, the company I founded and where I work now.

Having a good network of people who have the right skills, get to know you and your personality through mutual volunteer work, and are willing to help you is very important. I was lucky to have access to some really great people who helped me. I leaned on my network quite a lot, and I wouldn't be here without them. Every single person I know, from colleagues to personal friends, has done something to help me start this company.

As co-founder and CEO of Neurescence, Yasaman and her team develop fluorescence microscopy solutions for functional imaging of the brain and spinal cord, helping researchers better understand brain diseases such as Alzheimer's. But the opportunity did not simply appear all on this own. Yasaman took several proactive steps to make it happen, leveraging her network contacts and influencing the right decision makers.[12]

Although creating a successful startup may require the right opportunity, the stories of successful entrepreneurs demonstrate that an important part of winning the game is taking the initiative to explore new ideas, connecting with the people you meet along the way, and finding a way to create the opportunity you need.

Unique Strengths of the Scientist Entrepreneur

Most of the scientist-turned-entrepreneurs I've worked with credit much of their success to strengths that also make them a great scientist, either skills that they've learned or personal attributes that they've always had and probably lead them to become a scientist in the first place. Here are some of the strengths that are cited most often.

Problem solvers

Scientists are great problem solvers, and starting a business involves solving lots of problems. If you start your own company, you will be doing something that nobody has done before. People have started companies before, of course, but not a company to solve the specific problem you will solve. That means you will run into problems that no one has ever solved, and you will need to find your own solutions. The critical thinking and problem-solving strengths that you developed as a scientist will be very valuable.

> *I feel that becoming an entrepreneur was very synergistic with my research experience. One of the greatest things about a science PhD education is that you are constantly learning to solve problems you haven't even encountered before. That's incredibly resilient training and very useful for an entrepreneur.*
> – Peter S. Fiske, PhD in Geological and Environmental Sciences, founder of RAPT Industries[5]

> *The number one skill that most scientists have by nature is their critical thinking skills. They can look at a problem and figure out how to tackle it and arrive at a solution. No matter what field you're in, success is about solving problems and then moving forward and executing. That skill is what you're fine tuning as a scientist, and it is absolutely crucial in industry.*
> – Brit Berry–Pusey, PhD in Biomedical Physics, co-founder of Avenda Health[10]

Independent Learners

Scientists are independent learners. This skill is one of the hallmarks of a PhD degree, where we are given a problem that no one else has solved, and we have to figure out how to learn everything we need to become an expert. As Marinna described:[6]

> *I think many strengths that come from being a scientist are valuable for being an entrepreneur. For one, you're good at learning new things. The PhD degree is all about that. You take on a project that is on a brand-new topic that is so specialized that you didn't learn anything about it in your coursework. To be successful you have to figure out on your own what you need to know and where to learn it.*
>
> *That's definitely the case in the business world as well. As a scientist who becomes an entrepreneur, there is so much you need to learn about. Being independent, self-motivated, and able to learn things on your own is really helpful.*

Data Driven

The well-known and respected management consultant Peter Drucker said, "If you can't measure it, you can't improve it." You might therefore say that "If you want to improve it, you need to measure it." Scientists are great at determining what data to collect, how to collect data in an effective and reliable manner, how to analyze data, and how to determine what the data mean.

> *As scientists, we are trained in the skills of careful data analysis and the importance of data-driven conclusions that are independent from our*

own pre-formulated ideas. When speed is vital it can be tempting to make a quick decision believing that we intuitively know what is happening. Our experience as a scientist tells us to challenge that tendency and let the data speak for itself.
– Tanja Beshear, MS in Physics, Senior Quality Systems Manager at Medtronic[13]

Data-driven decisions are always valuable in business, but they are particularly useful when starting a company, because the process involves venturing far into the unknown. Again, from Marinna:[6]

The other thing that I've found really helpful is being data driven, because it helps to make many decisions. For example, when evaluating certain business deals, it comes naturally as a scientist to look up data on other similar deals to compare to the one you are considering.

The hard part about being a scientist in business is wanting everything to be heavily based on data. In the business world, things are less black and white. There's a lot more 'grey area' where the right answer is not so clear. Sometimes I find it difficult making decisions without lots of data to back them up.

Resourceful

Most scientists have learned how to make big things happen with limited resources. Research funding is challenging to secure, and so most scientists learn to be creative about getting their experiments done with little money. This practice pays off for the scientist entrepreneur:

When I first started my company, I needed to build a prototype but didn't have any money. I started emailing my contacts again and asked if they had any free samples I could have. They replied that they could send me out-of-spec parts, and I figured anything was better than nothing. Then they said they couldn't cover the shipping fee, and I said, 'That's fine. I can pay for shipping. I just can't pay thousands of dollars for parts.' So, my first prototype cost me some shipping fees and $5 for the 3D printing process.
– Yasaman Soudagar, PhD in Physics, co-founder and CEO of Neurescence[11]

Scientists typically spend many years as a graduate student making very little money. This teaches them to be thrifty and think carefully about how they spend their money in their private life. These habits are just what an early-stage technology company need.

In grad school, you learn how to live on $1500 a month, and after five years in grad school, you get pretty lean. Contrast that with these poor

business-school kids who work at Morgan Stanley for a couple years and then become entrepreneurs. After a couple years, they are having their shirts pressed every morning, and they've got their timeshare in the Hamptons. They become dependent on a high salary. Nobody goes into a science career with the idea of getting rich. You go into science because you are passionate about a subject and because you love learning, discovery, and the idea of creating something new. Those same passions are at the heart of entrepreneurship.
– Peter S. Fiske, PhD in Geological and Environmental Sciences, founder of RAPT industries[5]

Prone to challenging assumptions

If you are going to start a new company, you will need to do things that no one else has ever done before, and there won't be any formula for you to follow. While there is a lot of helpful startup advice available in books and articles and online courses, you will not be creating the same company that the people giving you that advice created. You will have your own unique product, your own unique customers, and your own unique team of human beings working with you to build a successful company. You will need to create your own game plan. That means that you need to think differently, question what you think you know, and question the advice others give you.

Scientists learn to question their assumptions early in their training:

One important skill you learn very early in experimental physics is to challenge your assumptions. You tend to make certain assumptions about your products and about the customer's application for your product.
– Tanja Beshear, MS in Physics, Senior Quality Systems Manager at Medtronic[13]

Not afraid of the unknown

As an entrepreneur, you will build a team of people and lead them on a quest. Many people are not very comfortable with the unknown, but scientists are. Science is all about venturing into the unknown, in a bold effort to explore it, understand it, and explain it. Scientists have chosen embracing the unknown as an essential part of their career, and this choice is well-suited for the startup world.

I find that if the problem we are trying to solve is something very tricky and high-level, where we are delving into the unknown, scientists have an edge. They often actually thrive on discovering the unknown.
– Ashok Balakrishnan, PhD in Physics, co-founder, CTO, and co-CEO of Enablence Technologies [14]

Interview excerpt: Ashok Balakrishnan on critical entrepreneur skills[14]

Ashok is a co-founder of Enablence Inc. and currently serves as the CTO and co-CEO. He holds a PhD in Physics from the University of Toronto. His full bio can be found in the Interviewee Bio section.

Dave: What would you consider your biggest career achievement?

Ashok: In 2004, a few colleagues and I formed our own company. The company is called Enablence, and that is where I'm working now. When we started the company, it was very small and just a way to stay employed and keep food on the table, but it has grown in surprising ways. Now it employs 25 people or so locally and close to 200 people internationally.

Dave: What would you say are the primary skills that enabled you to accomplish that achievement?

Ashok: The most important was just knowing what is important and what's not. Whether you are building a product or setting up a company, there are certain things that are important and have to be done right now. Other things may seem important, but don't really matter. It is critical to be able to assess a lot of pressures and then determine which ones are critical and which will not really matter in the end.

Being able to face the truth—whether it's something you like or not—is important. I learned this from graduate school. There were times in graduate school when I would struggle for months trying to figure out why I was not getting the results I was expecting. I would spend time pursuing problems I could fix more easily, such as mirror alignment or laser power, but all the time have this nagging feeling that something else was really the problem. It was something else that had been staring me in the face for a long time, but I was not willing to admit it. The moment you admit that, you can solve the problem.

Facing the truth is just as important in industry. Sometimes you are in the middle of developing a product and you realize, "Uh oh, this is not the way to go." You have to be honest with yourself and change direction quickly if you want to be successful, regardless of what you prefer. When you want to try to sell a particular business idea or product, you might think this is the niftiest thing in the world, but people may actually want something else. You have to deliver value, so be honest what that value is.

Unique Challenges for the Scientist Entrepreneur

While many of the strengths unique to a scientist can make us well-suited to launch our own business, there are also areas where some of our attributes and habits of thinking can present challenges. Here are a few situations where entrepreneurs find that the typical strengths and habits of a scientist might present challenges for them.

Can't always be the expert

We scientists tend to develop the habit of thinking that we need to find all of the answers ourselves, but this habit is often counterproductive for the scientist entrepreneur. Finding all of the answers was expected from us for our dissertation projects, and we tend to carry that habit into our private-sector careers.

This drive to find all of the answers on our own can be a real strength, but there are limitations. An important part of the playbook habit 'be effective, not smart' from Chapter 4 is relying on other people to help you and teach you some of what you need to know. This attitude is important for any private-sector career, but it is absolutely critical in a startup. One realization that entrepreneurs relay to me more often than anything else is that they simply cannot be the expert on everything, and they need to rely on another people to be successful. The required pace of progress is too fast, and the number of things they need to learn quickly is too great. As Yasaman Soudagar describes it:[11]

> *Persistence is good, but you have to learn to be efficient, and that is counter to the training many of us receive. In physics, if you get stuck on a problem, it is frowned upon to go to someone else for help. There is a sense that if you can't figure it on your own, you're not smart enough, so a physics PhD might spend weeks figuring something out, when someone else in their own group already knows how to do it.*
>
> *In industry, this is never appropriate. We are all one team, and we need to deliver a specific output at a specific time. We are like a ship where if something fails, we are all going to sink together. If you run into a problem, you need to get help from others right away rather than trying to figure it out yourself. That goes completely against what we learn as scientists, but we can't spend months doing a literature search the way you did when you were a student.*

Consider the case of scientist entrepreneur Chris Myatt, founder of LightDeck Diagnostics.[15] The device that Chris and his company developed diagnoses diseases by illuminating a patient blood sample in a specially designed cartridge with a laser to identify fluorescent labeled antibodies. This device provides a low-cost and portable alternative to the state-of-the-art approach, which requires sending blood samples to a lab with a flow cytometer, an expensive and time-intensive solution.

One realization that entrepreneurs relay to me more often than anything else is that they simply cannot be the expert on everything, and they need to rely on other people to be successful.

Chris is a physicist by training and brought a solid understanding of the photonics technologies required, but he had no experience with the science behind the antibody testing approach they would use. This meant that he could not even be an expert on all of the science used, much less the other important aspects of the device, such as the cartridge design and fabrication, antibody printing, and medical device qualification. Beyond the device itself, Chris needed to consider protecting the intellectual property they would create, financing options for raising the money they would need, human resources for staffing the team, and finding a location to house the new company. Attempting to have all of the answers himself would likely have resulted in failure.

Being a successful entrepreneur requires admitting upfront that you simply don't know everything and leaning on the expertise of others, and this often does not come naturally to a scientist.

There's no better place to plumb the depths of your incompetence than a startup. Every day you are saying, 'Oh, god! I don't know anything about that!'

– Peter S. Fiske, PhD in Geological and Environmental Sciences, Founder of RAPT industries[5]

Can't follow every curiosity

We scientists are naturally curious and tend to be driven to explore all aspects of the problem they're working on. We also tend to be thorough in the projects we complete, preferring to ensure that no aspect is unexplored or unexplained. The academic focus on certainty and complete understanding discussed in Chapter 2 reinforces this tendency during the scientists' training.

But a startup needs to be focused on achieving a very specific result, which is usually bringing their very first product to market and ensuring it is a success. Spending time on work that doesn't get them to the desired result is a waste of time and resources, and if taken too far, it can lead to the failure of the company. As such, a scientist entrepreneur needs to temper their natural enthusiasm for novelty and thoroughness, and instead identify and focus on exactly what matters.

This is related to playbook habit #2: successful scientists in industry figure out what matters and what doesn't. As an employee at a large company, a scientist likely has some time to learn this habit. As a founder, one needs to figure this out quickly and set an example for the team.

Infatuation with the technology

The most dangerous pitfall that I have seen many scientist entrepreneurs fall into is that they become enamored with their technology rather than identifying a real problem that their technology might solve. This is a natural risk for any of us who get excited by cool technology, but it's a real problem when an entrepreneur tricks themselves into thinking that there are customers who will buy their products when they see how great their technology is. I've seen many scientist entrepreneurs who convinced themselves that if they develop the technology enough, customers will see how useful it is and buy it.

But this is rarely the case. Potential customers you meet with may agree that the technology is really cool. They might even talk their managers into approving a purchase order to buy one unit to test it out and see how cool it is. But no good business executive will spend real money on a solution if it doesn't solve some business problem and increase the bottom line.

Consumer products are a bit different. We've all seen new technology that made a big enough splash that people were rushing to grab it. But even these products can struggle to 'cross the chasm'[16] into a market large enough to sustain the business. In the end, you need to make sure that your technology solves a real problem. Failure to do this will lead to the failure of the company.

I call this tendency to be enamored with the technology the 'technologist's siren song.' Just like the sailors in Greek mythology who crashed their ships on the rocky shores in search of the beautiful singing they heard, a scientist entrepreneur who pursues a novel and exciting technology without regard for the customer value proposition will crash their company on the rocky shores of the free market.

The right way to bring a product to market is to clearly identify a problem first and then develop the right technology to solve that problem. Too many founders start with the technology they believe has promise and then search for the problem that it will solve. This is backward and requires too much optimism. All too often, the result is failure.

I've heard several accounts of this problem from investors who have had bad experiences with tech-infatuated entrepreneurs who had no true problem to solve. The typical story is that the scientist founders over-emphasized their target customer's interest in their technology because they themselves are excited about it and did not sufficiently critique whether their product is solving a real need in the market. Less-experienced investors can be won over by a persuasive founder, resulting in months or even years of work and investment that is wasted when the product fails to achieve market success. Savvy investors recognize this tendency and will not invest if the founder hasn't clearly identified a real problem. These investors ask startup founders three important questions:

The Three Savvy Investor Questions:

1. What specific problem does your product/technology solve?

2. What are your target customers currently doing to solve this problem?
3. How much will your target customers pay to get your solution instead of their current solution?

If a founder cannot clearly answer these three questions, the investor will take their money elsewhere. Common responses from founders who have not fully identified the problem include the following:

Response 1: "We just need to develop the technology a bit more so that our customers will see how useful it is."

This response is generally the clearest indicator that the startup team is infatuated with the technology and has not identified a problem that they solve. In these cases, the team is hoping is that during the time they spend developing the technology to the point that it could be commercialized, they will be lucky enough to find the right customer with the right problem. This is rarely what actually happens.

Response 2: "Our customers don't currently have a solution to this problem. We will be their first solution."

This response often carries the insinuation that the startup will not have any competition and is therefore an attractive investment. Savvy investors know that this is not the ideal situation that the startup team believes it is. If their target customers do not currently have a solution, then they don't really have a problem to solve.

Response 3: "When our customers see the data we can give them, they will know what to do with it."

This indicates that the technology likely has application to the problems their customer solves, but that the team has not identified a clear connection between their technology and how their customer makes money. This is wishful thinking on the part of the startup team.

I actually spent several years in a company that had survived nearly a decade operating based on this statement. The company developed laser-based combustion monitoring equipment for large industrial furnaces. We knew that we could give them data showing them the oxygen and carbon monoxide concentration in different regions of their furnace, but we didn't know exactly what they would do to improve the combustion based on the data. We believed that when the furnace operators saw our data, they would make sense of it and make the proper adjustments. The company's investors believed us and continued to fund our technology development.

As it turned out, just giving our customers the data was not sufficient. We learned that we needed to work closely with them to figure out exactly how they could use the data to solve their problems. Since we were essentially developing a technology in search of a problem, there was no guarantee that we could achieve this. In some of the markets we pursued there was no problem to solve. We found that if they were not already solving the problem in some other way, there was indeed no problem worth spending money to solve. We were lucky enough to find that in two of the markets we were chasing, there was in fact a problem, and we were able to build a business in these markets. This experience was part of my own education and how important clearly identifying the problem is. I will not make that mistake again, and I now counsel new founders to be cautious as well.

Overcoming the Scientist Entrepreneur Challenges

What can a scientist do to make the most of their strengths, overcome the challenges, and improve their chances of success? Here are three suggestions I've collected from the stories of scientists-turned-entrepreneurs I've spoken with.

Surround yourself with good people

Resist the urge to try to have all the answers, and instead find people to help you. If you intend to win at the ultimate business game, make sure you build an expert team!

> *Early on, I thought that great technology—having a better invention or widget—was the key to a successful venture. I now appreciate that there are so many other things that are important, including having the right people around you. If you have a great technology and an imperfect team, you are in serious danger, but even an imperfect technology in the hands of the right team can be turned into an opportunity. That's the biggest lesson that I've learned, or more accurately, the lesson that I am still learning.*
>
> – Peter S. Fiske, PhD in Geological and Environmental Sciences, founder of RAPT industries[5]

Having the right people on your team helps you with the private-sector playbook habits outlined in Chapter 4.

Good teammates help you be effective and smart

Starting a company requires accomplishing many things quickly in a wide range of disciplines. You will also be faced with many tasks and projects that you have no experience with. For a scientist startup founder, this usually includes the finance and human resources areas and may also include manufacturing. Bringing on the right people with the right backgrounds will help a founder determine what matters and what doesn't in these unfamiliar areas.

It's very important to hire the right people. Building a company has a tremendous amount of hard work. I see technical people who fail at operating a business because they don't respect the knowledge base that is required. You have to learn those things yourself or you've got to put people around you who know what to do.

– Tom Baur, MS in Astrophysics, founder, CTO, and chair of the board of Meadowlark Optics[3]

It's helpful for a scientist entrepreneur to have a wide range of skills and be able to learn quickly, but there are obvious limitations. Trying to do everything will result in failure, so it's important to bring on the experts that you cannot be.

Here is how Yasaman Soudagar describes the importance of having the right team:[11]

When you start a company, you don't really know what you are doing. You just go forward and try things and see what happens. And what happens is that you make mistakes, and so you have to become extremely creative at finding ways around your mistakes.

You know how in the final years of your PhD you wake up every morning and think, 'Why am I doing this?' To get through it, you just keep thinking, 'One more day... one more day... one more day.' Well, I got to that point during the first year after I started my company, and I'd already gone through a PhD program. It is just that hard.

When you get to that point, you need to make a decision: 'Am I'm going to continue with this or am I going to quit?' And if you decide to continue, you have to do a lot of personal growth very quickly. Everything in industry moves so fast. You have competition, and they are moving really fast and you can't get left behind. That means you have to find your weaknesses and overcome them very quickly or else bring other people on to the team to help.

Good teammates help you figure out what matters

Starting a company involves making many decisions about what to pursue and what not to pursue. It's not always easy to know which tasks are the most important ones that will lead to the desired result.

Sometimes our own interests bias us toward the work we prefer to do rather than the work that really needs to be done. Sometimes the things we struggle with cause us to focus more on things that are easier to do. And as scientists, our curiosity and tendency to be thorough may lead us to spend too much time on things that don't really matter.

This is where having people with different strengths and interests who we trust to give us their objective opinions can be critical. As a team, you will be much better at deciding what matters and what doesn't.

Good teammates help you decide quickly

In Chapter 3, we discussed that making decisions when there is no right answer is a fundamental element of leadership. As the leader of a startup, it becomes critical to decide quickly when you don't have as much information as you might like to have. The team will act based on your decisions, and delaying a decision in the hopes of finding the 'right answer' can have significant consequences.

Make the phrase 'Make a decision, and then work to make your choice the right decision' a team mantra. Socialize this with your team and enlist their help in making quick decisions part of the company culture. This is an approach that helps many leaders be successful. We are leading a team into the unknown; indecision can be one of the worst things you can do. Not only do you not make progress, but the team begins to lose confidence. But as a startup founder, you've built a team of exceptional people. Make a decision, don't spend all your energy trying to anticipate hypothetical problems, and get the power of your team working to solve the real problems that come along.

People, not patents or publications, are the vector for innovation.

– Peter S. Fiske, PhD in Geological and Environmental Sciences, Founder of RAPT industries[17]

Focus on the problem

The advice to 'focus on the problem' has two different elements. The first is for the very early stages of your company, when you are deciding what you will pursue and what your product will be. The second aspect is for every stage of your company lifecycle when you are struggling to figure out what matters and what doesn't.

Early stage: Focus on the problem, not the technology

When starting your company, resist the tendency to be so excited by the potential of new technology that you aren't critical enough when evaluating the problems it might solve. Enlist your team to help ensure you are developing a solution and not just an exciting technology. Review the three savvy investor questions on page 97 with the team and ensure that you have clear, confident answers to all three.

The ideal way to develop a successful product, and a new company around that product, is to identify the problem first and then identify the technology that best addresses that problem. Of course, many successful companies are formed around a new technology that happens to find a suitable problem to solve. These companies are successful because the founders stay focused on the problem. They and the team are diligent about working with their target customers to clearly understand the problem and how their technology will solve it. Smart founders realize that customers may express excitement about their technology, but they will only buy a product if it brings them value.

Cellino, the company co-founded by Marinna Madrid, began with a technology in search of the right problem. They were successful because they met with customers to determine specifically what problems they might solve and then set up lab experiments to verify that their technology solved these problems.

Once a problem is identified, it is important the startup team aims their product solution directly at the needs of the customer. There is an identified risk in a startup spending a year or two developing the product that they believe satisfies the customer's problem only to release the product to the market and find that it misses the mark.

This risk of spending years developing a product only to find out that it is not actually the solution the customer really needs, is addressed in the great book *The Lean Startup*,[18] by Eric Ries. Eric proposes the minimum viable product (MVP)[19] as the best way to mitigate this risk. The MVP is the simplest version of the product that meets the most basic needs of the customer, and it is used to get feedback from early-adopter customers to guide future product development. The MVP is a great method to help the team figure out which features matter.

At each stage: Focus on the problem at hand

Once a startup has figured out what problem they are solving for what target customers, then their attention must shift to a new set of problems, focused on developing the product and delivering it to customers. As always, the team must figure out what matters. Successful founders find that if they identify the 'problem at hand,' this helps them focus on what really matters:

> *Once I jumped into business and realized the problem at hand was that a paying customer expected me to deliver something to them, it wasn't so hard to make that transition.*
>
> – Chris Myatt, PhD in Physics, founder and CEO of LightDeck Diagnostics[2]

Growth mindset

Things change quickly and frequently in a startup environment, and many things may not go the way the team expected. A founder needs to lead their team to embrace the challenges, embrace the failures and learn from them, and then move forward more intelligently and competently. A valuable perspective for helping founders and their teams achieve this is known as a 'growth mindset:'

> *In order to be successful working at any startup, you have to be really comfortable with the unknown and with making decisions without a lot of data. You also need to be really flexible, especially in the early stages, and you have to be really comfortable picking up new skills. Plans change, sometimes quickly and with little notice, and an employee has to be able and willing to adapt. Having a growth mindset is also very important, although that goes along with being willing to learn new skills quickly.*
> – Marinna Madrid, PhD in Physics, co-founder of Cellino Biotech[6]

The concept of a growth mindset comes from the book *Mindset: The New Psychology of Success*, written by psychologist Carol Dweck in 2006.[21] A growth mindset is an approach to work and life with the attitude that our strengths and weaknesses can be developed. Challenges present opportunities to develop skills, and failure makes us smarter.

The contrasting perspective is a 'fixed mindset,' which is the approach where one believes that their strengths and weaknesses are more or less determined for their entire life. You are either smart or not smart. You either have an innate ability or talent for something or you don't. When operating in this mindset, you perceive that you have a limited set of abilities, and so you'd better leverage those to the greatest extent possible.

You might say that in a fixed mindset, you believe you've been dealt a hand of cards, to use a poker analogy, and you have to play the hell out of the cards you have, because you aren't getting any more. People operating in this mindset tend to avoid risks and challenges, because failure might show others they don't have the abilities they thought they did. People operating in a fixed mindset also tend to dislike feedback, because they think it is pointing out that they don't have the right cards. Fixed mindset comes naturally to humans, and most of us operate in this attitude some of the time.

The sketch below shows a list of behavioral tendencies for the fixed mindset and growth mindset.[22]

FIXED MINDSET	GROWTH MINDSET
TRY TO LOOK SMART	WILLING TO LEARN
AVOID CHALLENGES	WELCOME CHALLENGES
FEAR FEEDBACK	ENCOURAGE FEEDBACK

Playbook Sketch 8 Behavioral tendencies of a fixed mindset and a growth mindset.

Note that the fixed-mindset tendencies sound very similar to the PhD stereotype of trying to appear smart. In a fixed mindset, one feels they need to appear smart to convince others they are worthy of their position. With a growth mindset, one recognizes that being willing to fail or be wrong provides the perfect opportunity to learn. Why spend your time looking smart when you could be getting smarter? When you are willing to admit you don't know all the answers, you put yourself in a much better position to get smarter. This is the essence of the playbook habit, 'be effective, not smart,' and it is an approach that many successful people have adopted:

> *I think everyone here gets tired of my quotes and sayings. There is a quote that I like from Steven Chu, who won a Nobel Prize in physics, developed laser tweezers for DNA, ran Berkley National Labs, and is now the U.S. Secretary of Energy. Someone once asked him what the secret was to making all these changes and doing them very successfully. He is reported to have said, 'When I go into something new, I make as many mistakes as fast as possible.' Here is a damn smart guy who is driven like there is no tomorrow, and he has the talent to go with it. He's saying that his secret is not that he avoided more mistakes than you, but he made more than you did and learned quickly from them. That's something I don't think very many people appreciate.*
>
> – Chris Myatt, PhD in Physics, founder and CEO of LightDeck Diagnostics[2]

When operating in a fixed mindset, one tends to avoid challenges, because they provide the opportunity to fail and appear incompetent. With a growth mindset, you recognize that challenges are how we grow, learn, and develop into better people. When you are willing to embrace opportunities to fail, you put yourself in a much better position to grow stronger.

With a fixed mindset, one is afraid to receive feedback from others, because feedback is seen as highlighting one's failures. The founder who operates in a growth mindset embraces feedback from others. They see it as an opportunity to observe one's performance through their coworkers' eyes and identify and solve issues that they may not have otherwise seen. Rather than avoiding feedback and the development opportunities that it offers, one embraces feedback as a gift from coworkers and an opportunity to get better.

Here is how Marinna described the importance of a growth mindset for her team at Cellino:[6]

> *Growth mindset is something that has always been important to us, and we emphasize it a lot with the team. Our team is really small, and so everyone has to wear a lot of hats. A new responsibility might feel uncomfortable at first, but the right person is OK with that and keeps working to develop the new skills that make them better in the new role. If there were someone on our team who didn't have a growth mindset*

and wasn't willing to try on new hats, it would be so hard for the team to be productive.

My own role has changed so much since we began Cellino. I find that every few months I'm doing something different. I've spent months doing nanofabrication in the cleanroom, I've spent months doing biology experiments at the bench, I've spent months writing patents, and I've spent months interacting with potential customers. If I didn't have a growth mindset and wasn't willing to pick up these new skills, it would be very stressful for me and my team members, and we wouldn't have made nearly the progress that we have.

Interview excerpt: Marinna Madrid on the importance of the problem[6]

Marinna is a PhD physicist and co-founder of Cellino, a biotechnology company that is using lasers, robotics, and machine learning to build a fully automated approach to cell engineering. She and her lab partner, Nabiha Saklayen, co-founded Cellino in 2017 after discovering that their research had valuable applications in synthetic biology. Marinna's full bio can be found in the Interviewee Bio section.

Marinna: The correct way to build a business is to identify the problem first and then figure out the best possible solution for that problem. But when you're creating a startup out of a PhD project, you are working backwards. You're starting with the solution and then trying to find the right problem that your solution can solve.

Dave: I've seen that with so many entrepreneurs coming out of universities. They have a technology and are in search of a problem. That's a tough hill to climb.

Marinna: Yes, it sure is. You spend a lot of time just trying to find the right market fit and the right application. And once we found the application we were really interested in, the work that we had done during our PhDs was not even necessarily the perfect solution; so as a result, the technology that we've developed at Cellino has actually evolved far beyond what Nabiha and I were doing during our PhD days.

Play the Ultimate Game!

Before I close this chapter on startups, I want to mention my favorite book on entrepreneurship, *The Hard Thing about Hard Things*, by Ben Horowitz.[23] This book is a compilation of Ben's blog posts relaying lessons and telling stories from his own experience with entrepreneurship. Most are based on Ben's experiences building the enterprise software company Opsware that Ben and Marc Andreessen founded in Sunnyvale, California in 1999 and sold to Hewlett Packard in 2007.

The great thing about Ben's book is that he acknowledges up front that the biggest challenges in growing a business or things for which there is no formula. This is summed up in this excerpt from the introduction:

> "Every time I read a management or self-help book, I find myself saying, 'that's fine, but that wasn't really the hard thing about the situation. . . .'
>
> "The problem with these books is that they attempt to provide a recipe for challenges that have no recipes. There's no recipe for really complicated, dynamic situations. There's no recipe for building a high-tech company; there's no recipe for leading a group of people out of trouble; there is no recipe for making a series of hit songs; there's no recipe for playing NFL quarterback; there's no recipe for running for president; and there's no recipe for motivating teams when your business has turned to crap. That's the hard thing about hard things—there is no formula for dealing with them."

This is a great description of what I call the private-sector game. There's no formula, no recipe, no checklist that will tell you everything you need to know. This is true as an employee in the private sector, and it is even more true for the entrepreneur. Success requires taking risks, making decisions when there is no right answer, and finding your own way to win the game.

Starting a company is not for everyone—but if you would like to see your research make an impact within a few years; if you have an appetite for taking risks; and if you believe you possess the flexibility, decision-making abilities, and growth mindset that will allow you to navigate the startup world, consider starting your own venture. It's a lot of work, but as other scientists who have started their own companies will tell you, it is a very rewarding game to play!

> *What I'm doing is very tough, and I know this sounds cliché, but I don't feel like I'm working. I know many people who are just doing a job and they're miserable. To me, this is completely different, and I think it comes from my start as a scientist. When you're pursuing a PhD, you're doing it to follow your heart, right? Let's face it: with the salaries in academia, no one is doing it for the money, and that takes your career to a whole new level. It's not just a job, it is something that it is fundamentally satisfying. When you complete a PhD, you experience that joy of doing something for your heart. If you can find a similar thing in industry—which is possible; some people think it's not possible, but it is—you will continue that experience.*
>
> – Yasaman Soudagar, PhD in Physics, co-founder and CEO of Neurescence[11]

The fact that Avenda Health is in a position to potentially cure someone with prostate cancer is so rewarding. We just treated our first clinical patient with our product last week, and just knowing that we developed something from a concept all the way through to FDA clearance and now it's been used in humans to treat prostate cancer is really, really rewarding. Even more rewarding is the idea that, because we are a small team, everything about getting to this milestone has my hands on it.

One in seven men in the U.S. is diagnosed with prostate cancer, so if our technology could help one in seven men, the impact of that is just mind-boggling. It's very rewarding knowing that my contribution may help the lives of all those men who are fathers, sons, and grandfathers, but also the lives of all those people they love, whose lives are also affected by the disease.

I was not expecting the emotions that came with seeing that first case of our product hopefully helping someone get rid of their cancer. It's a really great feeling knowing that all the hard work, the stress, the sleepless nights, and the gray hair that I now have, has all come together to potentially save lives. I will say right now that I'm just having so much fun and I've also never been so stressed in my life. Those things are not mutually exclusive.

– Brit Berry–Pusey, PhD in Biomedical Physics, co-founder of Avenda Health[10]

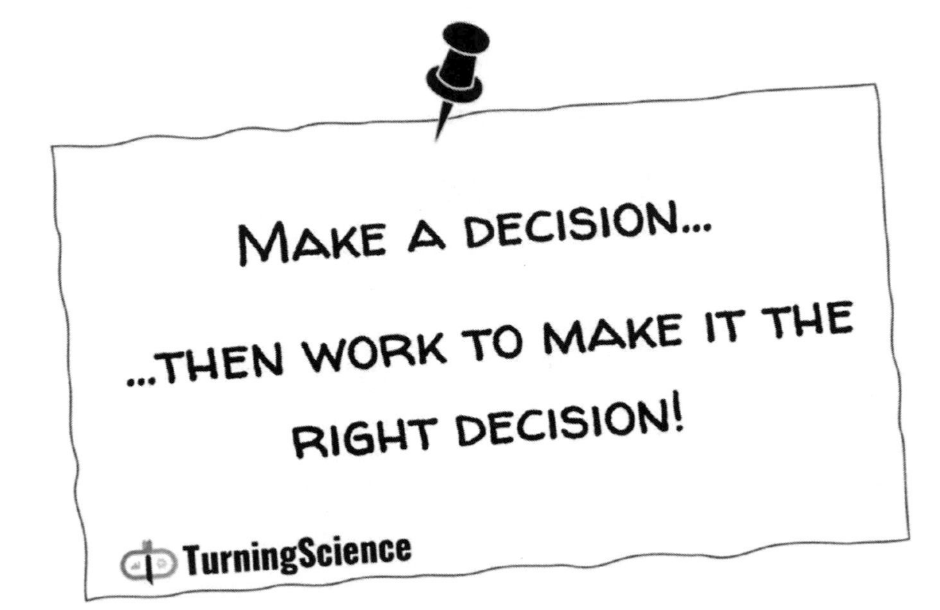
MAKE A DECISION...
...THEN WORK TO MAKE IT THE
RIGHT DECISION!
TurningScience

Chapter 7
Your Career Is a Game

Some pro tips for your private-sector career playbook

Throughout this book, I have described the private sector as a place that is best approached as a game, because of these three important principles:

There is more than one way to win,

Winning requires taking risks,

Knowledge alone does not make you successful.

Since being successful working in the private sector requires learning to play the game, it should come as no surprise that finding a job and building a career in the private sector is also best approached as a game. If you feel that your career is not everything you expected or dreamed it would be, perhaps you need to work on your game!

I am frequently approached by early-career scientists who asked me, "What additional training should I get in order to get the job I really want?" From this question it is clear that they are still taking the formula approach to their careers. Asking what additional qualifications they can add to their resume so the hiring manager will see them as the best choice for the job amounts to looking for the 'right answer.' I tell them that if they have a PhD, they already have more education and training than most people on the planet. I tell them they need to stop thinking in terms of 'qualifications' and learn to play the game instead. I say, 'You already have plenty of qualifications for what you want to do. Rather than trying to add more credentials to your resume, figure out what you want to do, find people who are doing it, and convince them you can do it also!'

In this chapter, I relay five private-sector playbook 'pro tips' to help you improve your own career, supported with stories from many of the industry scientists I've interviewed. I hope will give you new ideas and new inspiration for your career planning.

Pro Tip #1: Start with Why

I pulled the title of this pro tip from Simon Sinek's excellent book of the same title,[1] because I believe that starting with the purpose behind why you are doing something is one of the most important elements of success in just about anything you pursue. Scientists who describe the most rewarding private-sector careers can describe what it is that they set out to do with their lives and careers, and they know just how the career they've designed fits that purpose.

If you would like to build a more meaningful career than you currently have, I suggest you start by identifying your 'why.' To help with this, I've started this chapter with the 'why' stories of several successful industry scientists.

> *I distinctly remember that as a kid there were two things I wanted to do when I grew up. One was to be a stockbroker so I could wear fancy shoes. The second was to be a doctor so I could help people. I am not a physician or a stockbroker now, but I am a businesswoman who uses science to help people. And so, when I look back on my life, I realize that I'm doing exactly what I wanted to do as a child.*
> – Brit Berry–Pusey, PhD in Biomedical Physics, co-founder of Avenda Health[2]

Why work in the private sector?

The career paths of the industry scientists I've spoken to are very diverse, as are the reasons they love their careers. But when I look at the reasons why they choose to pursue a career in the private sector in the first place, two common themes emerge.

Theme 1: Didn't want the lifestyle of a professor

After years of observing the professors we worked with as graduate students, many scientists—myself included—decided that that we did not want their career or lifestyle. This may have been related to the work–life balance we desired, or it may have been that we weren't interested in a career spent chasing research funding, but either way, we wanted a different life and career than what we had observed.

Theme 2: Wanted to make a bigger impact

Many scientists felt drawn to make a larger impact than we believed we would achieve through publishing our research in academic journals that would be read by a handful of people. We wanted to make things that would change many people's lives in a year rather than in two decades, so we decided to go turn science into things people need today.

I feel like I'm on the path to make a huge impact with my work. There is a song by Ariel in the Disney movie The Little Mermaid that has the line: 'I don't know when, I don't know how, but I know something's starting right now.' I often hear that line in my head. There have been other moments in my life when I had that same feeling, and what I was doing at the time turned out to be very impactful. I have that same feeling right now.

– Sona Hosseini, PhD in Engineering; Applied Science, Research, and Instrument Scientist at Jet Propulsion Lab[3]

Interview excerpt: Roger McGowan on the professor lifestyle[4]

Roger McGowan is a PhD physicist and Sr. Research Fellow at Boston Scientific Corporation. His full bio can be found in the Interviewee Bio section.

Dave: What motivated you to pursue a career in industry?

Roger: By the end of my postdoc, I knew I wasn't interested in the lifestyle of a university professor. It's hard to achieve a good work–life balance when you have to be an instructor, a researcher, and a proposal writer all at the same time.

I knew that I would like to teach someday, but I wanted to have a solid background in industry first. Many physics students will end up in industry, yet most professors don't have an industrial background. I feel that if I want to be a really good teacher, I should first have that industry experience so I can help my students make the transition.

Another reason I went into industry is the money. I knew I was going to raise a family and wanted the higher salary typically found in industry. Finally, I liked the idea of creating something useful.

Dave: How has your perspective changed since you first went into industry? Is there anything you would do differently with the knowledge you have today?

Roger: One thing I would have done differently is to plan for going into industry while I was in grad school. I value what I learned from my postdoc, but I would skip it if I had it to do over again. Now I realize that three years gaining experience at a company would have been far more meaningful than the three years I spent as a postdoc.

When I first went into industry, I envisioned myself ultimately reaching a director-level position and guiding a team to achieve a common vision for our technology. My experience since then has shown me that a director of technology doesn't necessarily deal much with technology. I wouldn't want a director-level job at the large company I'm with now because they spend their time pushing paper and dealing with budgets.

Dave: What have been the most rewarding aspects of your career to date? What excites you or keeps you motivated in your career?

Roger: I enjoy working in the medical industry. Making devices to help people and save lives can be very rewarding. I also enjoy the mentoring and the teaching opportunities that come with leading teams and projects.

As for what keeps me motivated, that would be a desire to create something and to have an impact on the company. I have a strong internal drive to develop myself, exceed the expectations of my manager and team members, and consistently achieve larger and larger opportunities.

Interview excerpt: Christina C. C. Willis on staying in the lab[5]

Christina is a laser scientist with a PhD in Optics from the College of Optics and Photonics (CREOL) at the University of Central Florida. Her full bio can be found in the Interviewee Bio section.

Dave: What drew you to science as a career?

Christina: For science, there wasn't a single moment. I enjoyed science from an early age, particularly my science classes, and I had several science teachers who were influential for me. But I can tell you with a fair amount of certainty that I became a laser scientist because of the movie *Real Genius*, which is one of my favorite movies. I started watching it around age 12, and I've seen it well over a hundred times. The film was made in 1985 and features a precocious 15-year-old boy who gets into a fictitious version of Caltech and works with an eccentric senior student on a high-power-laser project. I started studying optics as an undergraduate, and when I went to graduate school, I sought out a research group where I could specifically work on high-power-laser development. I'm sure there is a strong correlation between those two things.

Dave: That's a great story! So, then, at some point you decided to transition into industry?

Christina: Throughout my undergraduate studies and entering graduate school, I was just following the 'traditional' career path to a faculty position without questioning it. That just seemed like the obvious answer. However, there were things I observed along the way that slowly changed my mind about the career I wanted.

As an undergraduate, I had an advisor who periodically became challenging to interact with, and I slowly realized that the shift in his demeanor happened when grant proposals were due. I watched him cycle several times through that process of stress. It didn't look like something I wanted to do. That was the first data point for me that perhaps a career in academia wasn't the right fit for me.

Then I went to graduate school, and I noticed that my advisor and many other successful professors didn't spend much time in the laboratory doing experiments. I realized that when you are a professor running a large research group, it doesn't make sense to spend a lot of time in the lab because you're supposed to be doing the high-level things like figuring out where your group is headed next, coming up with visionary ideas, and making sure that everyone gets paid. It's not really possible to do all that and the nitty-gritty of experiments at the same time.

I didn't like the idea of working myself out of the lab. I really enjoy using my hands and building things, and so that was the final decider for me. If I was going to continue working on lasers, I wanted to be doing the actual assembly and testing, so I decided to look for industry jobs after graduate school.

I ended up working at two different industry jobs, both of which were small government contractors. The first was a startup, and I think I was the 13th person to come on board. The second was larger, around 100 people.

Dave: Many scientists experience a real culture change going from a lab (where the goal is to generate new knowledge) to a company (where the goal is to produce something for a paying customer). Was this a big shift for you?

Christina: No, not too much. A lot of the work I did at my first job was similar to my graduate work because we offered measurement, software, and custom design services rather than an off-the-shelf product.

It was similar at my second company, which specialized in the design of custom laser solutions for specific applications. I think it was a very different atmosphere than it would have been at a company that makes hundreds or thousands of a given product. I also spent more time on R&D end of the work, doing more proof-of-concept projects rather than building deliverable systems. So that was also similar to the lab work I did in graduate school.

Interview excerpt: Oliver Wueseke on what it takes to be a professor[6]

Oliver (Olli) Wueseke has a PhD in Molecular Biology from the Max Planck Institute for Molecular Cell Biology and Genetics in Dresden, Germany. His full bio can be found in the Interviewee Bio section.

Dave: Let's start at the beginning. What led you to a career in science?

Olli: I was always fascinated with nature since I was about six years old. That fascination grew into a passion for biology after discovering genetic engineering during my late high school years. The more I exposed myself with the material, the more enticing I found it, leading me to pursue a PhD in biology. I'm fascinated by the complexity of biology and how to make sense of it. I love the challenge of going into the unknown and charting new territory.

Initially, I had this vision of becoming a professor and teaching people all of the exciting things I had learned. Things changed with time, and I did not become a science professor, but my fascination with biology hasn't changed.

Dave: How did you end up transitioning to industry instead of becoming a professor?

In my PhD, I focused on studying protein interactions and building *in vitro* systems to investigate complex protein dynamics. As I neared the completion of my PhD, I spent much time thinking about what I would do next, wondering if I should pursue a path in industry. For me, recreating life in a petri dish is fascinating research. Still, it wasn't clear how my research would translate into an industry career.

After interviewing for a biotech investment fund in Germany, I decided I wasn't done with science yet and chose to do a postdoc. I went on to do my postdoc with brain organoids, and after two years, I met one of the founders of System1 Biosciences in Vienna. He asked me if I wanted to come to San Francisco and build their brain organoid platform. I would do the same work but in a startup with lots more money and a lot more people. So that's what I did. That was my transition to industry, and it was terrific.

When I set out to do my PhD, I was pretty adamant about becoming a professor. Then I realized that there are many PhD students and not very many professors. My initial response was that I would just have to work harder. Dealing with this was probably the hardest challenge I've ever had in my life. Still, I eventually came to understand that I had other strengths that would help me be successful in different career paths. It turned out that there were several elements of the career of a professor that conflict with who I am. It was vital for me to understand who I am, what I'm good at, and how I bring value to other people. This knowledge has made it so much easier to determine the path that I want to follow. I guess you could call this life planning, all starting with a large amount of honest self-reflection.

Interview excerpt: Kerstin Schierle–Arndt on wanting to do big things in industry[7]

Kerstin is a PhD chemist and Vice President of Research Inorganic Materials and Synthesis with BASF. Her full bio can be found in the Interviewee Bio section at the end of this book.

Dave: Where did your interest in science begin?

Kerstin: I would say it started when I took chemistry in school. I'm a very logical and curious person and so I enjoyed science right away. I also liked the interplay between theory and experiment, where we could study a theory and then do a practical demonstration in the lab. When I went to university it was obvious to me that I would study chemistry.

Dave: How did you decide to start your career in industry?

Kerstin: I honestly never thought about staying in academic research. For me, it was always clear I would like to work in industry, because I always knew I wanted to do big things. A research career at university stays involved with tiny little laboratory-sized quantities in most cases. I really wanted to see the things I do in big quantities, like in trucks on the road and train cars on the railway. That was a big motivation for me to go to industry.

Dave: That's convenient that you knew early on what you were going to do.

Kerstin: Yes, although if you would have asked me as a young student, I think I would not have been able to articulate my intentions. Somehow, I had a picture in my mind of someone who is doing science research that ends up in a product.

My advisor during my PhD work was a very good mentor and he helped me to start my career. He had several collaborations with the chemical company BASF, so I had opportunities to meet the scientists who worked there from time to time. There was also a course for PhD students that exposed them to BASF, and he wrote a letter of recommendation so that I got the chance to take part in that course. The course was very interesting and helped me understand clearly what industry was all about. The combination of fundamental science research in my advisor's lab along with his connections to industry helped me figure out what I wanted to do with my career.

Dave: Did you make the transition directly from the University of Bonn to BASF?

Kerstin: Yes, although the fact that that this worked out so seamlessly was to some extent by accident. I had the opportunity to take an internship at BASF and work in their laboratories for two months during my PhD thesis. It was very elucidating and also introduced me to the person who would later become my boss. I also got some insight into when the next permanent positions would be available. It turned out there was an ideal position for my background, and the timing was perfect for when I would be finishing my PhD thesis. I decided to apply and got the job.

Interview excerpt: Jason Ensher on making a bigger impact[8]

Jason has a PhD in Physics from the University of Colorado at Boulder. His full bio can be found in the Interviewee Bio section.

Dave: Can you tell me about the science phase of your career?

Jason: I developed an interest in optics and quantum mechanics by my senior year of college, so I applied to the graduate program at the CU Boulder because it had a large number of faculty in atomic, optical, and laser physics. What really won me over was a visit to the school. It seemed to have a great program, and I liked the fact that Eric Cornell and Carl Weiman were working to achieve Bose–Einstein condensation (BEC). I was also interested in living in Boulder. Eric was building his group at the time and needed some help over the summer, so I started working in his lab right away, studying ultra-cold atoms. By June of 1995, we had succeeded in achieving BEC. After I graduated, I decided to take a postdoc position at the University of Connecticut where I worked on creating ultra-cold molecules out of ultra-cold atoms.

Dave: So after an excellent start in academia working on research that earned your advisor the Nobel Prize, you decided to go into industry. Can you describe what was behind this decision?

Jason: During my postdoc, I was starting to feel a little dissatisfied. I was thinking that maybe a tenure-track position was not going to be fulfilling for me. I was doing interesting work, but it didn't feel like I was reaching enough people with it. I wanted to do something that would have an impact on a lot of people, not just the several dozen colleagues around the world who might be reading my papers. Also, working with Eric and Carl in graduate school, I learned how to be successful in research, but I saw the investment that it requires. I decided that I was not willing to commit at that level. I wanted a little bit more out of the rest of my life. At the same time, fellow students who had graduated were reporting general satisfaction working in industry. I remember receiving an email from a friend who was working at 3M, and he indicated he had discovered this remarkable thing called Saturday. In the labs in graduate school, we typically worked much of the day on Saturday. Industry also provided higher compensation and a tangible reward because you produce things that people use.

Through our work on BEC, I learned that doing something new and risky can really pay off. I took that mentality with me to industry. I liked the idea of a smaller company that was charging ahead into the unknown doing something novel and interesting. My first job was at ILX Lightwave. Telecommunications was growing rapidly at the time, and so was ILX. They had just opened a small development lab in Boulder, and there were only two other PhD physicists there, so it seemed like a place I could carve a niche.

Dave: What have been the most rewarding aspects of your career?

Jason: I have found it rewarding to work with teams of people and to learn the leadership and organizational skills necessary to direct the activities of a team. I've also enjoyed the opportunity to work in so many different areas and learn about so many different things. I've worked at four different companies in a little under ten years—not by choice, as I was laid off twice—and so I've worked in telecommunications, test and measurement hardware, optical component manufacturing, remote sensing, aerospace, and holographic storage. It's been rewarding to keep my career going and yet do different and interesting things.

Author's note: The value of a publication

I have a scientist friend who I hike with frequently, and we discuss a wide range of topics about life and our careers. He shows a real interest in the work that I do to help scientists develop rewarding industry careers, and one day we were discussing why we decided to build our own careers in the private sector.

One day, my friend told me a story about his days as a grad student and spending months working on a journal publication only to realize that all of the work had very limited value:

> *When I was getting my PhD in Materials Science, there was a definite moment when I realized my pursuit of an academic career was a waste of time. I had been working on a piece of research that I was trying to get published. After a couple rewrites and two journal submissions I finally got it accepted for publication. I was at the lab getting ready to head home when I got the email from the editor saying my article was accepted and attached was a final proof. It was getting late, and I was hungry, so I printed the final proof, headed to my car, and started driving home with the article on the passenger seat. I had to stop for gas, and as I was reaching for my wallet, I saw the paper and suddenly I had a revelation. This paper that I had spent so much time on was literally worthless. I could not exchange it for gas, rent, or anything else. Also, since it was university research I did not own any of the intellectual property. I didn't even own the copyright of the article that I wrote! From then on, I knew I needed to follow a different path.*

After having spent more than 20 years in the private sector, I'd become very used to thinking about the financial value of the work I do. My friend's story really grabbed my attention. Science is tremendously valuable! How could a science publication be of so little value?

We scientists all know that journal publications do indeed have significant value to a limited audience, but at the time we submit them, there is a very narrow selection of people who appreciate that value. Some of us struggle with this realization. When we choose to work in the private sector, we see that our work is usually appreciated by thousands of people, if not millions or tens of millions, depending on the product and market we serve. Many of us decide that this is a better fit for our own personal career interests. It's a better fit for our 'why.'

Pro Tip #2: Understand Your Strengths

When I mentor scientists on designing a rewarding career path, many express the feeling that the private sector holds so many different options that they don't know where to start. Some say that it feels like being in the middle of an ocean with water in all directions, and they are trying to decide which direction to swim to find land. How do they pick a direction? I always encourage them to start with their strengths—what they do well and what they don't do well. Understanding your strengths is a critical step in finding a career path that will be exciting and rewarding.

What can a scientist bring to the business world?

Scientists are trained in a laboratory setting. We are taught to investigate the unknown and find explanations for phenomena we observe. We collect data, analyze it, and suggest hypotheses that support our observations. We learn to solve complex problems in very systematic ways.

But these skills don't sound like a very good description of what is needed in the private sector, where products and services are developed and sold to customers. What can a scientist contribute to the world of industry, aside from the obvious option of working in a research and development lab? The following interview excerpts describe several valuable strengths that scientists can bring to their industry careers.

Interview excerpt: Christina C. C. Willis on the value of persistence[5]

Dave: **What are some of the strengths that have been important for getting you to where you are in your career now?**

Christina: The first one that comes to mind is persistence. My PhD dissertation was a really hard time for me. I had what felt like an unending sequence of things break or go wrong with my experimental setup. I realized that if I wanted to finish my PhD, I had to learn how to just get up, keep going after each setback, and learn to let go of my expectations regarding schedules and specific outcomes. Simply refusing to give up is the only reason I finished.

Another important skill that I acquired during my PhD is an ability to learn new things. The ability to not get frustrated when something is hard and to have the willingness to say 'I don't know this stuff, but I'm going to put in the work to figure it out' is a big part of that. I think that ability is correlated with persistence, because the more you get frustrated, the more you're going to want to give up and the harder it is to persist.

Dave: **Have you found that strength to be useful even after graduate school?**

Christina: Graduate school was certainly the hardest point in my career so far, but the persistence that I learned has been very useful. Certain problems just don't get solved unless you refuse to give up. The ability to stick with a problem I don't know how to solve, a problem where I have no idea what's wrong and maybe all I want to do is just run away from it—that has been valuable.

Interview excerpt: Roger McGowan on the value of learning and teaching[4]

Dave: What is the biggest achievement in your career to date?

Roger: When I first joined Boston Scientific, I worked in a process development team designing a laser bonding process for their stent delivery catheters. The existing process had been optimized by varying parameters until the desired result was achieved, but the team didn't understand why it made a good catheter. The new process I helped develop had many controls and monitors so we could observe the effect of varying each parameter as we were making a catheter. It was essentially an engineering project, but what made it truly successful was my emphasis on a fundamental scientific understanding of the laser weld.

This project was successful enough that the new process was extended to all of our next-generation laser systems. Championing this effort led to significant career development for me within the company. Showing them the value of a scientific approach led to the creation of my current position, Research Fellow for Process Development. Being successful in this effort is what I'm most proud of in my career.

Dave: That is a great demonstration of the benefit of using a scientific approach in a product design setting. What would you say are the top two or three skills that enabled this achievement?

Roger: The most important thing was a solid technical understanding and the ability to teach myself. These are skills that I learned in graduate school. They gave me the confidence to jump in and quickly learn about CO_2 lasers and other things that were new to me. Without these skills, I wouldn't have had the ability to demonstrate how my ideas would improve the process.

Another important skill was my experience in teaching and mentoring. Much of my success resulted from presentations that impressed senior management. It was critical to be able to teach them and help them understand why my ideas were of value.

My ability to dissect a problem was also very important. Any technology brought to market will have numerous challenges along the way. You have to be efficient and effective at leading a team to dig in and work through these problems.

Interview excerpt: Oliver Wueseke on atypical scientist strengths[6]

Dave: What are some of the strengths that have been instrumental in building your career?

Olli: There are a few that have been very important. The first is inspiring people through a vision. I seem to be able to take complex goals, formulate them into a vision, and communicate them in a way that people understand and see it as achievable.

Second, people have told me that in times of adversity, being an empathic leader and building a safe zone for the team to operate in is something that I do well. This ability requires creating an environment of trust and being vulnerable as a leader. A team with a high level of trust and empathy can pull off some fantastic stunts even in the face of severe adversity. As an example, think of astronauts, soldiers, and firefighters. They achieve what they can because a high level of trust creates enormous amounts of resilience.

Humor is another strength that I'm very proud of, and it's a subtle skill that has a very positive influence on team dynamics. I always think that if I take myself too seriously, it will be my downfall. Being able to laugh about myself allows me to be vulnerable, keeps me level-headed, and can take the heat out of intense situations.

Dave: Those are not strengths one might expect from a typical scientist.

Olli: I do have some of the classic scientist strengths, such as being a critical thinker, being open-minded, and loving the challenge of solving a hard problem. However, I think that vision and empathy are especially valuable when leading a team into the unknown. They help me to give people the trust and confidence that we can tackle this challenge and that they are operating in a protected space. As a team, we venture out and try new things every day. We have the resilience to recover from all the setbacks that come with doing science because we trust and appreciate each other.

The value of 'non-technical' strengths

When we work in a laboratory environment, our technical strengths are generally the most important. These are generally the strengths that we developed and cultivated during our time as a graduate student. We are probably most familiar with what these are, because we spent a lot of time developing them and being critiqued on them.

But most career paths in the private sector involve a lot of activity that is not purely technical. These might include your ability to speak and write persuasively or your ability to work with and influence others. You may need to explain a technical instrument to a non-technical customer and find yourself needing a whole new vocabulary to do so. You may be part of a product design team and need to describe use cases that outline how your customer may use your technology. You might even find yourself helping to develop a product roadmap, where you envision what new technology developments your company will need to develop in the next 3–5 years. Do you have the strengths that will make these tasks enjoyable and allow you to be successful?

As you get further into your career, you are likely to find that the strengths that helped you the most early in your career become less important, and there are new strengths that make you successful. This is a good change and is part of staying relevant as your career progresses. Technical skills are harder to maintain outside of a hardcore research environment, and it's also hard to continue competing with younger scientists coming out of universities. Changing your career focus is a wise way to stay relevant, but that involves developing new skills and new knowledge. Often the strengths that become increasingly valuable as your career advances are non-technical strengths.

Some scientists find it easy to identify their non-technical strengths, and others struggle with the question. My interview with Oliver Wueseke addressed how he identified strengths that were instrumental in helping him design a rewarding career.

Interview excerpt: Oliver Wueseke on determining your non-technical strengths[6]

Dave: How did you get in touch with who you are and what you are good at?

Olli: What helped me was to do things outside of my typical science job. As a PhD student, I was an active student representative. I learned that I couldn't stand doing event planning. All this phone calling, back-and-forth communication, keeping track of everything, and merely executing a plan was exhausting. However, I also found my passion for inspiring others when I had the chance to give a motivational and emotional speech to the whole institute. During my postdoc, I started reviewing and writing scientific papers for friends and colleagues. I also gave workshops on how to implement storytelling in your scientific presentations. I just tried it, and people found it beneficial.

These interactions with other scientists taught me a lot about the additional skills I have. Most aren't part of the natural science training, but I picked them up along the way and found them very useful in almost all settings. These were skills that I seemed to be good at and that I enjoyed using. They helped other people and were valuable to me in terms of making money. I only discovered them because I had ventured into new things and offered help to other people. Most of this I did in my free time, but I have to admit that my lab work also suffered a bit. I didn't care; I was doing the things I loved.

Dave: It sounds like that was a good trade.

Olli: Understanding these things about yourself will ensure that you can choose a more meaningful career path. Start by asking yourself what you like about science. Do you like writing papers, or do you hate it? Do you like working in the lab, or do you only like discussing science with smart people? Any answers here are correct, but there are plenty of jobs where you can discuss science with smart people and don't have to write.

Leverage the people around you to try out your ideas, test what's working, and get honest feedback about what they think your strengths are. Sometimes you find that who you think you are and who others think you are is not the same, and that can be a problem. I was lucky to have lots of good people around me who guided me in the process.

I think this issue of understanding who you are and what you are good is valuable for any career path. But how do you find out if you are good at something that you have never done? You don't! For example, I only learned the difference between a project manager, product manager, program manager, and line manager after leaving academia. The differences are significant, but I was never even exposed to the concepts in academia. Only by trying things out did I realize that I like product management but that I'm not a good project manager. I also learned that I love creating a vision and a strategy and leading people to achieve it.

You have to commit to action to understand what you like and what you do well. I suggest people make an active effort to seek these things out. Reading and thinking aren't enough. You can't simply think about it and read a book about how you do something new. You have to get out and try it. We teach PhD students to engage a challenge by researching information in papers, books, and at conferences; and then to think deeply about a possible hypothesis and come up with the appropriate experiment to test the hypothesis. However, most real-world experiences do not require rigorous hypothesis testing. When I tried to learn to play drums, I did not read about the physics of how a drumstick could potentially create different sounds when I hit the drum in different spots. I took the drumstick and beat the drum, over and over again. By doing so, I learned a lot about playing the drums. Acquiring the same amount of knowledge through vigorous hypothesis testing probably would have taken me a lifetime. I encourage everyone to stop hypothesizing and seek the learning experience instead. Keep the hypothesizing for doing science at your job.

Dave: Yes, I have also found great value in understanding who I am and what I'm good at when planning my career. It's led to some of the most exciting work of my life, including what I'm doing now with TurningScience.

Olli: I couldn't agree more! And the first steps to hopping on this path are quite simple. Ask yourself the following: In the past week, what were the tasks I enjoyed doing and why, and what were the tasks that I couldn't stand doing? I call this energy boost & energy drain analysis. Asking and answering the *why* is critical here because it will help you identify the underlying aspect of a specific task that gives or drains your energy. To me, this is the quickest path to find your real strengths and weaknesses.

Dave: I like your approach of focusing on strengths and interests. It seems that formal education is so often focused on improving weaknesses, as though there is some skill set template that we all need to match in order to be successful.

Olli: There is a discussion about leadership that asks whether it's better to foster strengths or try to boost weaknesses. In my opinion, the latter just gets you to a mediocre level in everything you do. I think that improving your strengths is probably the more valuable path.

Interview excerpt: Brit Berry–Pusey on evolving strengths[2]

Brit has a PhD in biomedical physics from UCLA and is the co-founder and COO of Avenda Health in Santa Monica, California. Her full bio can be found in the Interviewee Bio section.

Dave: Have you found that as your career progresses, the strengths that are most valuable to your career evolve as well?

Brit: Yes, definitely. There are skill sets that you start developing in graduate school that you develop and rely on throughout your career no matter what stage you're at. For example, one of my big strengths is a confidence that I can figure out whatever problem is thrown at me. That was very important in graduate school, and it is still very important. But now I spend much more of my time thinking about how I present myself and how I communicate to others. I also spend my time on new tactical skills such as how to manage budgets and manage projects to ensure that things happen on time. I also have to think about how to tend to the emotional needs of the employees in my company so that I make sure that they're getting the experiences and learning opportunities that are valuable to them and still performing to the level that we need them to.

Pro Tip #3: Create Your Own Opportunities

It's easy to look at someone else who has been successful and assume that they were just lucky enough to come across the right opportunity. But in my experience, most people who find success went looking for the right opportunity and played a very active part in finding it or even creating an opportunity where none existed.

If you can envision the opportunity that would allow you to make the impact you want to make, why wait for the right opportunity to come along? Perhaps you can find a way to create the opportunity you want, and others will assume that you are just lucky.

People are always blaming their circumstances for what they are. I don't believe in circumstances. The people who get on in the world are the people who get up and look for the circumstances they want, and if they can't find them, make them.

– George Bernard Shaw

Interview excerpt: Oliver Wueseke on creating opportunities[6]

Olli: When people ask how I found that (first industry) opportunity, I have usually said that I was just at the right place at the right time, but I don't think that's correct. I've observed that successful people often create opportunities for themselves. They create the possibility to be in the right place with the right people at the right time. I realize now that's what I had done as well.

Dave: I've always liked the quote 'Luck is the residue of design.'[9] I've observed that people who are seen as 'lucky' put themselves in the positions where they come across great opportunities and can take advantage of them.

Olli: I agree, but that second step where they seize the opportunity when it comes along is essential. I think that's the hardest bit, at least for me. I often find myself contemplating whether or not I should do something. Then time passes, and then the opportunity is gone. Generating opportunities is excellent, but you have to commit to action to appear 'lucky.' In hindsight, I think many of the things that seemed like luck for me were the result of creating an opportunity and then being ready and willing to act when it came along.

Dave: Your point about 'doing something' is very relevant for those of us that have a lot of science training. We are trained to be thinkers, and in the science lab we see that more data and more analysis lead to better results. But outside of the science lab, thinking and analysis become an excuse for delaying action.

Olli: I fully agree with that assessment. It's a classic scientist's problem to think we should read a bit more about a problem before we do anything. It's such a natural temptation, and it's what we trained to do, so it feels like the most comfortable path out. In reality, often, it's not.

Interview excerpt: Christina C. C. Willis on creating opportunties[5]

Dave: You've had a very interesting career path so far. So, now you are in the middle of a one-year congressional fellowship, working on science policy for the U.S. government?

Christina: Yes, though I'll draw a distinction between 'science policy' and 'science for policy.' 'Science policy' involves creating legislation that regulates the conduct of science, and 'science *for* policy' is about using science to ensure that legislation is based on sound principles. In my fellowship, I work more on the 'science for policy' side, leveraging my scientific background and my ability to do research and data analysis.

Of course, the irony of all this is that I left academia so that I could stay in the lab, and yet now I've stepped away from laboratory work, and even from working directly on science. I use many of the skills that I acquired from my science education, but my policy portfolio has little to do with science itself. But I haven't yet given up on the idea of going back to lab at some point.

Dave: How did you make the transition from industry to this fellowship?

Christina: I really enjoyed both of the industry jobs I had, but I also recognized that I wasn't maximizing my writing and communication skills in those roles. My shift from being an industry scientist into public policy was based on an interest in public policy and wanting to explore those communication skills in more depth.

I've always enjoyed explaining things to other people, and I have this general urge to make sure no one is confused. I was first exposed to a policy environment through 'Congressional Visit Days,' which I started doing as a graduate student and continued after graduating. These were events where we would talk to congressional staff and explain what optics was, why it matters, and what their member or office could do to support optics and fundamental science. I *really loved* getting to explain complex scientific topics to people who had little science background. These are very capable, intelligent people, but they typically haven't studied a lot of science. So, it requires a different way of communicating than when you're talking to another scientist.

It's so exciting to see the person's eyes light up when they understand what you are saying. It was so invigorating to think that I could make a positive impact on science funding and policy and engage in the excitement of communicating at the same time. The policy fellowship provided an opportunity for that.

Dave: That's very interesting. So, you arrived where you are now by realizing you had a strength that you were not using and matched it with an interest you had identified in your past, and then made a career change to match them up.

Christina: Yes, absolutely.

Dave: That's a very valuable career and life design tactic! Are you finding it rewarding?

Christina: Yes, the fellowship has been incredibly rewarding, and getting to see how a congressional office works is an amazing learning experience. That's something few of us get to see, and yet what happens there impacts us all.

Pro Tip #4: Think Big!

When I give workshops for PhD students who are nearing graduation and planning a career in the private sector, one of the things I teach is a series of steps for building their own rewarding career path. The last one of the steps is "think big!" This last step is meant as an inspiration, for the students to think beyond the confines of what they have likely been told are the career possibilities for a scientist.

I want to extend that inspiration to any scientist at any stage of their career who is considering their private sector career path. Scientists can do so much more than research! One of the great aspects of it being a game is that anyone can play. That doesn't mean that you will always win, of course, but you can get out there and compete at just about anything you want to do.

In previous generations, there was a lot more focus on sticking with a traditional career path. People would fit themselves into a career because they had to feed the family. In this scenario, your university education would define you for a large portion of your life. The good news is, times have changed, and our current and future generations have more flexibility than the previous ones. Instead of saying, 'I am a person who works in this particular job,' I can say, 'I'm a unique person who brings these unique skill sets and experiences.' With this perspective, the more different environments you have worked in, the more valuable you are. It's in line with the mindset of lifelong learning. Career paths have become more flexible, and it's just natural that your

> *coursework will become irrelevant within a few years of work. However, that's okay. You will gain experiences that are more valuable than the textbook education you got at the university.*
> – Oliver Wueseke, PhD in Molecular Biology, founder and CEO of Impulse Science[6]

But the one big problem with thinking big is that you can only think about things that you can imagine. That's why it's so important that you expose yourself to many different ideas so that you have more options that you can imagine for yourself. Read about different career possibilities. Consider things that you think would be exciting, even if you can't imagine a scientist doing it. Perhaps you would make your dent in the universe by bringing your own unique set of strengths to that career.

Think like a designer

When you think about designing a career path that would be exciting and rewarding for you, think like a designer does. As Dave Evans and Bill Burnett point out in their great book *Designing your Life*, "Designers imagine things that don't yet exist, and then they build them. And then the world changes."[10]

This is what you are hoping to do by designing a better career path. You imagine something that doesn't exist yet, at least for yourself, and then you build it. And in doing so, your world will change.[11]

In the introduction to *Designing your Life,* Evans and Burnett have a section called 'Think Like a Designer,' in which the first two steps are 'be curious' and 'try stuff.' I completely agree with this approach, and so the first two sections under this section of 'think big' will follow their recommendations.

Seek new ideas and perspectives

When designing your career path, be curious and look for new ideas and perspectives. Think creatively. Talk to people and get new ideas. Read the stories of other scientists in industry.

Try new things

When designing your career, you need to get out and try things. New ideas and perspectives are wonderful and useful, but they are not enough. Like any game, you can't figure out if you enjoy it or if you are any good at it by sitting at your desk and reading or thinking about it. You have to get out, talk to people, try things, and figure out what you're good at and what you're not, as well as what you would enjoy and what you would not.

> *The data to make good decisions are found in the real world, and prototyping is the best way to engage that world and get the data you need to move forward.*
> – Dave Evans and Bill Burnett in Designing Your Life[12]

Get outside your comfort zone

One thing that scientists who have successfully moved into new career paths appreciate is the growth opportunities provided by trying things that stretch you and may make you uncomfortable. Having the courage to try something that may seem scary or causes a fear of failure is an important part of growth.

Collect your own data

While there is great value in getting ideas and advice from others, what worked for them might not work for you. Their stories of success and failure are an excellent way to gather new ideas and tactics and add them to your own playbook, but you have your own unique strengths and interests. It is important that you decide what works for you.

> *The career plans we initially have usually come from trying to copy what other people have done. The problem with that is, you only really gain traction on that path if it aligns with your natural strengths and interests. For instance, if I have all the strengths and skills that would make me a good professor, it's easy for me to copy the paths of other successful professors. However, I don't think that every person that starts a PhD naturally has the strengths and skills it takes to be a professor. If they try to copy that path, they'll fight themselves and probably end up in a miserable state after a while. I think that happens quite often with PhDs and postdocs in the life sciences. We also can't blame anyone for it; it's merely that in this field, the path toward a professorship seems the only natural progression path. We're missing alternative role models, which is why this book is so important. So before attempting to copy someone else's career, I would start with a little self-reflection. Try to understand who you are, and then make sure that who you are overlaps with what you want to do. Ensuring this congruence makes your career planning so much easier. Pick your job based on who you are, what your strengths are, and what you want to do, rather than picking a career path that someone else followed and trying to make it work.*
>
> – Oliver Wueseke, PhD in Molecular Biology, founder and CEO of Impulse Science[6]

One of the most damaging things for scientists who are building careers in the private sector is to rely too much on input from professors and others who have built their entire careers in academia. Most of these people are very well-meaning and want to help out, but because most of them have only worked in academia, they cannot offer much constructive advice about a private-sector career. What's worse is that there tends to be an attitude in many universities that academia is the one true career path for a scientist. Even a well-meaning professor is likely to be influenced by this attitude, and that influence is likely to impact the advice they give to anyone considering an industry career. Sometimes the impact is minor, but sometimes it is very damaging.

Don't listen to the people who tell you that you can't achieve your dream. They are speaking from their own ignorance or their own fears and limitations. Those fears and limitations are not yours. Think big! And think for yourself.

Interview excerpt: Sona Hosseini on being a scholar[3]

Sona has an M.S. and PhD in Engineering Applied Science from the University of California, Davis. Her full bio can be found in the Interviewee Bio section.

Dave: What would you tell a young scientist who has decided they want to make an impact?

Sona: If I could say one thing, I would say, 'When you finish high school, stop acting like a student and start becoming a scholar both in your career and in your life.' I wrote a column in our college newspaper my freshman year discussing this transition. I went to college in Iran, because my family moved back to Iran when I was 12 years old. The English translation of the column title was something like 'One who used to learn knowledge transitions to one who seeks knowledge.'

In Farsi, there are two different words for a student before they enter college and after they enter college. The word for a student before college is dāneshāmuz (دانش آموز), which translates as 'a person who learns knowledge.' The word for a student in college is dāneshjū (دانشجو), which means 'a person who seeks knowledge.' In English, we use the word 'student' for both, so that distinction is not clearly made.

In Persian culture, when you go to college you transition from being taught ideas put forth by others to a position where it is now your responsibility to seek knowledge. If you just read books and take final exams, then you are missing the point. The intern college students that I work with here in the U.S. still seem to be in the role of learning knowledge. When I tell them about the two different terms in Farsi, they don't seem to identify with the idea.

Interview excerpt: Kate Bechtel on 'banging around'[13]

Kate has a PhD in Physical & Analytical Chemistry from Stanford and is a Biophotonics Fellow at Triple Ring Technologies in the San Francisco Bay area. Her full bio can be found in the Interviewee Bio section at the end of this book.

Dave: Are you a planner when it comes to your career?

Kate: It's funny you ask that. When I was finishing my PhD, I had cold feet about whether this was the life I really wanted. I was a bit terrified, as every probably graduate student is, about what to do with the rest of my life. As I mentioned, I'd always thought I'd be a professor, and I viewed life as this super-straight line, where every single choice leads to the next thing, and there's no deviation. Once I choose a path, I'm stuck there forever. That's what I thought my career would be like.

And I remember confiding in one of my PhD committee members that I was petrified about my career, worried that whatever choice I made would constrain me for the rest of my life. How could I ever make a choice like that? He looked at me like I was crazy, and he said, "Just bang around a little. Life isn't deterministic, it's stochastic. Just choose something and go with it. If you don't like it, go do something else." My mind was completely blown realizing that I actually had the freedom to change my career direction if I didn't like what I was doing. Suddenly, it all seemed less frightening.

Dave: How has the stochastic model worked for you? Where has 'banging around' led you?

Kate: I've benefited greatly from trying different opportunities that have come along, and it's really helped me figure out what I like to do best. I've been a technical lead for a project, and I've been a project lead, which is more of a project management role—I was a functional manager with a team of direct reports—and then I was a Practice Area lead for the biomedical optics area, meaning I was responsible for all the business and the revenue stream for that area. I've also worked in business development, helping to land new projects and seeking out new leads. I was very fortunate to get to try all of these different things, and I found after doing all of that, I really liked being a technical lead the best. I just really enjoy analyzing data and figuring out what it means.

I have a colleague who started at Triple Ring about the same time I did, and we both worked our way up to being Practice Area Leads at the same time. It was then that I decided I had too many different things to focus on, so I chose to go back to being a technical lead. My colleague, who had been following the same path, wanted just the opposite. He wanted to do more of the functional and practice area work. We worked it out to combine our groups, and I gave him all the management and administrative work, and I took all of the technical work. Now we are both really happy with what we are doing. And we only got to this great place by 'banging around' and trying different things to see what we liked and what we didn't like.

I think most people can find opportunities to try different aspects of a career in technology, and if you can identify what part of your job you love the most and that makes you the happiest, you should stick with that.

Dave: We scientists learn that we can describe things with a formula and get the right answer by collecting more data and doing more analysis. I find there's a tendency to think that we can figure out our career paths the same way.

Kate: That's right, but just thinking about it doesn't give you an authentic view of the career. You have to experience it. It's like a roller coaster: You can sit and watch people on a roller coaster screaming in happiness or in fear, but you don't know how you're going to behave unless you try it. No amount of studying the equations and knowing how much force is being applied to you in that little car during the ride will help you figure out what your own response will be.

Interview excerpt: Christina C. C. Willis on leaving her comfort zone[5]

Dave: I'd love to hear more about your year in Japan! How did that opportunity come about?

Christina: The summer before my senior year of college, I realized that I was tired of studying and I wanted to take a break from being a student. I was working as a Summer Undergraduate Research Fellow, a SURFer, at NIST in Gaithersburg, MD. When the laboratory director, Dr. Katharine Gebbie, offered to make time to meet with the summer students, I sought her out to ask for her advice. I told her that I was planning to go to graduate school but that I wanted to take a break and travel first. She recommended that if I wanted to go to graduate school I should keep working in science, because if I didn't it would be incredibly difficult to come back to school at a graduate level.

That was how the idea of working abroad came about, but the specific opportunity in Japan was not planned at all; it was Dr. Gebbie's suggestion [. . .] It was an unexpected opportunity that came to me through networking, and I had the sense to grab it. I'm so happy that I saw that and pursued it.

Dave: That's a really creative solution that allowed you to meet your personal needs without risking your career progress.

Christina: It was certainly not an easy decision; I was very afraid, at the age of twenty-one, to move to a country where I didn't speak the language or know a single person. But I recognized that several important contacts had been references for me to give me this opportunity, and I would be letting them down if I didn't take it. I knew that ultimately it was a really amazing opportunity for me and that I would also be letting myself down if I didn't take it.

I'm so glad that I decided to go, and it informed an important part of my personal philosophy: that growth happens outside your comfort zone, that you don't grow or learn new things unless you're challenged. I didn't recognize going in how much of an impact that year in Japan would have on my personal life, but I did recognize that working at a national lab in Japan would be a great career move for me.

Dave: Can you share any stories about how you came to this realization about growth and your comfort zone?

Christina: Yes, I can. While I have pretty good verbal skills and I am generally good at explaining things to people, I never had any formal training on presentation skills before I got to graduate school. I was very fortunate to have a large research group that regularly held presentation reviews and I learned an immense amount from those sessions. Before anyone gave a formal presentation, they were expected to present it to the group and the group always gave very thorough feedback on all aspects of presentation. It was always supportive, but it could also be quite unvarnished.

When I did my first practice run of my proposal presentation, I felt like it got completely shredded by the group, and my initial response was to feel depressed. It was hard for me to hear about all the things that were wrong with my presentation and how much work I had to do to fix it. My lab mates weren't being unkind by pointing out its shortcomings; quite the opposite, they had given me their time and detailed feedback. That sort of feedback was very educational, it made my presentations better, and it really improved my speaking skills. While it was sometimes challenging and uncomfortable, I was fortunate to have a research group with experienced members who were willing to take their time to give each other constructive criticism so directly.

Dave: So, the practice of putting yourself out there to receive uncomfortable feedback was the key to valuable growth. That's a great realization. I've found it's often hard for those of us with lots of education to be open to growth opportunities like this. We learn that we need to be smart and that we succeed when we are right, so it's painful to be wrong. But being open to growth opportunities, even when it's uncomfortable—and perhaps *especially* when it's uncomfortable—is incredibly important.

Christina: Yes, it really *is* important. It's not smarts alone that have gotten me to where I am in my career; the strengths that I've developed around being persistent, consistent, adaptable, and willing to accept feedback have played an invaluable role.

And this discussion reminds me of a concept that I discovered through my yoga practice. Back when I started doing yoga, I had a friend who was significantly more advanced than I was. I went with him to an advanced class that was ultimately beyond my capabilities. During that class, I realized that I had two choices: get frustrated and give up, or simply accept my failures as part of the learning process and keep trying. I realized that being the least able yogi in the room gave me the most space and impetus for improvement.

Recognizing that you're not the smartest or most accomplished person in the room often offers a similar opportunity. Do you leave the room because you are afraid to fail or look foolish, or do you decide to stay and use it as an opportunity for rapid growth and learning?

You come to realize that you grow when you're outside of your comfort zone, trying new things, or acquiring new skills. You realize that when you are not afraid to put yourself in a difficult situation or when you allow yourself to be surrounded by people who are better at something than you, that's when you will grow the most. And you realize that being willing to be bad at something, being willing to struggle and to not punish yourself for struggling or even failing, is itself a very important skill.

Dave: Is there anything you tell yourself in the moment when facing a challenge to help you with the right mental attitude?

Christina: There are two things that help me. First, I remind myself that if it were easy, someone else would have figured it out already. This helps me remember that there is value in the persistence required to accomplish a difficult task.

Second, I have a personal analogy that helps me process the discomfort often associated with growth. I was once struck by a car while riding a bicycle. A significant portion of my treatment was deep tissue massage to address the whiplash caused by the accident. It was very therapeutic and helped my neck, but my body's impulse in response to the deep pressure was to tense up, which would have prevented the massage therapist from giving me an effective treatment. I had to learn to relax into that intensity to receive the full benefit of the therapy. In much the same way, when I have a hard problem to solve, I try to relax into the discomfort instead of letting frustration distract or disrupt me.

I have also found that taking breaks when my brain gets saturated is important. When I hit a point of diminishing return, I do something else, like having a cup of tea, going for a short walk, or practicing yoga and meditation to give myself a reset, after which I can resume working on the problem productively. Yoga and meditation are also excellent practices for developing self-awareness and being able to recognize that moment of saturation.

Interview excerpt: Sona Hosseini on collecting your own data[3]

Dave: Has there been anyone who made a big impact on you and your career in a very short time?

Sona: During graduate school, I was really excited to go to NASA and pursue novel science questions that required developing new instruments for space missions, but I was also excited to see what I could do for commercial applications. I was in a workshop where Anita, one of the very famous scientists in the field, was also there. I managed to talk to her during the first poster session, and I started passionately talking about how maybe after graduate school I wanted to look at industry options. She looked at me and carelessly said, 'The best people stay in academia and become professors, and the people who cannot become professors go to industry.'

I remember feeling like I was this blossoming flower, passionately talking about all my ideas, and when she said that, I crumpled. Her opinions mattered a lot to me, and I just lost my confidence. I stayed away from her for the entire week of the workshop. Someone I had cared so much about had just told me my entire vision was wrong and I was on the loser's path. I usually had lots of fun during workshops, but not that week.

Dave: That must have been very discouraging. How did you recover?

Sona: I was depressed for a few weeks, but then I remembered what my grandpa had encouraged me to do several years earlier. Just like when I started taking college classes in high school to see what subjects I enjoyed, I started visiting different work environments to see what I enjoyed. At the time, I was working on my PhD project at Lick Observatory on Mount Hamilton, so I would be up there for a week or two at a time and then drive down the mountain to see friends and dance Argentine tango till late at night. Almost the entire tango community in the Bay Area works at tech companies, so I had lots of opportunities to talk with my friends from both industry and academic career paths and explore my own feelings about their situations. I talked to friends from Stanford University, the University of California, Berkeley, NASA Ames Research Center, and a number of different companies. As we'd talk about each environment, I observed my feelings to see what might fit me best. In the end, I decided I wanted to go to JPL, not because of what Anita and others told me, but because of my own data.

When Anita told me that losers go to industry, I felt uneasy for weeks. Collecting my own data made a big difference for me and allowed me to be much more confident in my decision. During my last year in graduate school, I had couple of other people tell me the same thing, but I was no longer uncomfortable. By that time, I had gathered my data and done my homework and I didn't crumple. I did come back stronger, but it took time and some planning. I know so many other people who would not come back up, and I can only imagine how many other people she told this to. What's important is that we follow our own path, not the path of others. We can only succeed if we are following our own stars.

Dave: So, you took an experience that might crush some people, and instead you used it as a growth opportunity to strengthen your confidence in the path that you chose.

Sona: I always see conflict as a moment of growth. If I hit a barrier, I'm either going to break it or build a ladder over it, or I may come to the conclusion that the barrier served me well because it stopped me! In any case, it's always a moment of growth.

Imagine if I had ignored Anita, if I had said, 'Screw her, she doesn't know me!' What about the next time I heard that same opinion? After all, maybe she was right. Maybe she had given me a gift by telling me how things really work. I needed to gather my own data. Now that I've done that, every time the conversation comes up, I know what I'm talking about and I'm no longer uncomfortable. It's all about the data!

Every university I've attended, the message was that the proper end goal is to become a faculty member. I think it's very irresponsible for professors to train all of their students to go into academia. Professors train their students to write papers and do proposals, but students are not usually trained to communicate well with others or even themselves. In industry, communication with others is so important, and we end up disadvantaged.

We scientists have been trained to gather data and make decisions based on data. But there are so many situations in life where we need to make a decision without being able to gather our own data, and it's completely wrong to make decisions based on other people's data. This is the situation when we consider moving from academia to industry. If we look to our professors and their data, their success leads us to conclude that the best career is in academia. We have an obligation to do the research and look for our own data. If you are being told that academia is the only good path and you shouldn't go to industry, you as a researcher need to go get data for yourself.

Can a scientist change fields mid-career?

I speak to many scientists who are interested in switching to a different field than the one they are currently working in, or perhaps the one they studied as a graduate student. Many are hesitant to consider this switch as a real possibility. While many university science departments are now embracing cross-disciplinary degrees and research projects, many other scientists still tend to think of themselves as suited for working only in the discipline, or the perhaps even the specific narrow research area where they were trained and have been working since graduation.

> *Scientists tend to identify with their discipline and see themselves only as a chemist or physicist. They lock themselves into very specific roles within a company instead of thinking, 'I am a problem solver. I am resourceful. I can do a broader range of things.'*
>
> – Peter S. Fiske, PhD in Geological and Environmental Sciences, Executive Director for the National Alliance for Water Innovation[14]

But many scientists do successfully change to new areas and even move into a completely new discipline. The limitation for the scientists who are hesitant to attempt a change seems to be that they see their strengths in the knowledge that they have obtained through their postgraduate studies and through their work. I encourage them to think more broadly about the real value of their PhD. It turns out that many employers are not so focused on the domain-specific knowledge that we've gained through years of study and work. Many are more interested in the strengths we possess that allowed us to learn all that knowledge so quickly. Here's how Kate describes it:[13]

> *We see a lot of PhD scientists who've been trained to focus heavily on the details of their dissertation work and highlight something unique that they have learned. But that's really not the best approach for getting a job in the real world. What's best is to present skills and how you are able to solve problems for the company.*
>
> *At my company, we've hired people who worked on science in grad school that we have no explicit need for. We've hired a solar physicist, for example. We have no need for solar physics, per se, but we knew that this person was able to understand a challenging topic at a deep level, understand sophisticated instrumentation, interpret the data and determine what is real and what comes from the instrument artifacts, and make decisions on how to move forward. Those are the types of skills that we need in a PhD scientist.*
>
> *The most important value of a PhD is the ability to teach yourself and become knowledgeable and useful in a short period of time.*

If you think you'd like to move into a new area, the fact that you may not currently be an expert in that new discipline does not prevent you from making the transition. Technical knowledge is not the only important thing.

When we seek to take a career in a new direction, it's easy to worry that we will not be able to compete for a position in a completely different technical discipline. We imagine that there will be plenty of other scientists who have just the right match for the technical skills the company is looking for. But this is not always the case. Experienced industry managers know that there are many things that are important when bringing in a new employee. Good managers know that and technical skills and knowledge are generally not the most important. Much more important are non-technical strengths such as work ethic, creativity, critical thinking, communication skills, and the right personality to fit within the team dynamics.

We PhD scientists have proven that we can quickly and independently learn to do something completely new and have the persistence and the problem-solving abilities to be successful in that new field. If you want to take your career into a new area, don't worry about not having the depth of knowledge that someone already working in that area has. You can learn what you need quickly and bring your personal strengths along. You did it as a graduate student, and you can do it again.

Interview excerpt: Scott Sternberg discusses moving to a new discipline[15]

Scott has an MS in physics and is the Executive Director of the Boulder Economic Council in Boulder, Colorado. At the time of our interview, he was the Executive Vice President of Services at Vaisala. His full bio can be found in the Interviewee Bio section.

Dave: How did you transition from laser spectroscopy in a physics lab to a neurobiology lab?

Scott: I decided that I wanted to take my skills, and everything I'd learned, and move into a new application area. After I completed my Master's, I took a research position in the Anatomy and Neurobiology department at Colorado State University. There was a large focus on spinal cord research at the time. Spinal cord neurons typically don't regenerate when damaged, meaning that if you break your back, you can't walk again. We did an experiment where we were able to block certain chemical processes that inhibit regeneration and demonstrated millimeters of nerve cell growth. It was a small step, but orders of magnitude improvement over the microns of growth that had been demonstrated previously. The experiment was heavily reliant on optical technology and software digital image processing that I helped develop using my physics and optics background. To bring a skill set I had developed in physics research and make a significant contribution in a new environment was very rewarding. It was also an affirmation that my career did not need to be confined to one particular path or application area.

Dave: How did you make the transition into industry?

Scott: One of the companies I was working with on the nerve regrowth project was Photometrics in Tucson, Arizona. They were making digital imaging systems for telescopes. We were taking these CCD cameras and putting them on microscopes. Instead of pointing the cameras at outer space, we were pointing them at inner space—the physics was almost identical. After buying my third camera and "actively persuading" them to rewire and reprogram them, they finally said, 'You know, you are the kind of person we want in our company.' It was a wonderful transition to do something completely different and move into a business environment.

My transition was into an application specialist role where I was dealing with scientists. Our customers were scientists, so for me it was a good impedance match between the business and the academic environments. Academic customers don't necessarily like talking to traditional sales or businesspeople. They prefer to talk to someone who understands the research world they are working in. If you have a science-research background, then you have the 'street cred' they are looking for. I used to call myself a translator because I would translate academic speak into business speak, and vice versa, and help information flow as efficiently as possible.

I consider making this transition to be my biggest accomplishment. At that point I recognized that my ability to understand the interests of both the business and science communities was a unique skill that I could bring as an employee. From that point on, I have worked in business environments that are heavily entrenched in scientific innovations or scientific research.

Interview excerpt: Yasaman Soudagar on the importance of culture fit[16]

Yasaman has a PhD in physics from the École Polytechnique de Montréal and is the co-founder and CEO of Neurescence in Toronto, Ontario. Her full bio can be found in the Interviewee Bio section.

Dave: Scientists often wonder how to find a job with a company that fits them and their strengths. What do you look for when you're looking for someone to join your company?

Yasaman: It is extremely valuable to have the right person for the job, and that means two things: a person who fits the culture of the company, and someone with the right expertise. You can imagine how difficult it is for a company to find someone who is a match for both of those things. It's like a puzzle, and there are very few people in the world who will fit.

Dave: How do you find someone with the right fit?

Yasaman: It is very, very hard to find someone who will match the culture and who also has the exact technical expertise you'd like to have. If you can find that person who is a good match for both, that is golden. We insist on a culture fit but are flexible on technical expertise.

Dave: Can you say more about that?

Yasaman: A startup is very different than a larger company, and so the culture we look for is really different and really important.

We look for people who are comfortable taking the lead on a project, because we are a small team and there's a ton of work that needs done. I don't have the time to manage everything they do, so they need to be independent.

We also need people who are flexible and able to adapt to change very quickly. During an interview, I always tell the person, 'Today, this is the big project that we are hiring you for, and that is your top priority. But on day three of your job, I'm going to come in and tell everybody the priorities have changed. Everyone needs to stop what you're doing and start on this other project [. . .] and you need to be able to do that.'

Also, because we are a small team, our employees end up covering a lot of different roles. For example, if you are hired to help with a technical project, you will not work solely on that technical project. Tomorrow, we may start a collaboration, and you will be sitting in meetings with a new team defining a new project. The day after that, I may suddenly need you to collect some numbers for a business report.

We also have to work really fast at times. I may send you a bunch of papers one morning and need you to give me two PowerPoint slides by the end of the day that answer three specific questions in a way that I can present to investors tomorrow. These are the kinds of deadlines I deal with, and I need your help to distill the information quickly and make it presentable. This is just the nature of a startup. Some people just cannot work like that. We hire people who can.

I'm brutally honest when I'm interviewing people, and I tell them exactly what it is like to work in this environment. I have had people who come back to me after their job interview and say, 'You know, I thought about this, and I really don't think I can work like that.' I really appreciate that honesty.

> So, if we find a person who is a really good match for the culture and we think they are going to be able to work in a startup environment, then we don't need the technical expertise to be a perfect match. You can always train the person. If they are in optics and you are doing something optical, then that's not a big leap, right? The same thing goes for someone trained in neuroscience. The much bigger issue is the culture piece.

Pro Tip #5: Fortune Favors the Bold!

As a final thought about building a rewarding career in the private sector, find the courage to take bold action. Building an exciting and rewarding career is a game. As such, it is not achieved by learning the right steps, collecting the right credentials, and following the rules better than anyone else. It is achieved by setting your sights on the goal and going after it with bold action. If you don't make it the first time, try again.

In pursuing this goal, it is natural to doubt ourselves. I used to think that people I worked with who had already achieved so much in their careers must be very confident and know exactly what they are doing. I've learned since then that this is not true. Most people who are successful still doubt themselves, just like you and I do, but they keep going and don't let either the doubt or the obstacles stop them. They don't let uncertainty and failure stop them from achieving their goals. Here is helpful wisdom from Brit Berry–Pusey related to this point:[2]

> *When we talk about imposter syndrome, it's important to remember that most of us have this challenge. It's easy to think that we are the only ones who feel like they're faking it, but when you realize that most people feel this way at some point in their lives, it brings a sense of comfort.*
>
> *When I'm feeling self-doubt, I remind myself, 'There is nothing more that I love than a challenge,' and I find that feeds me. I recall that I've proven myself many times before, and I'll do it again. I recall that I do have the ability to tackle challenges and that I actually end up enjoying the struggle. This practice motivates me to try harder and to do better.*

The other important thing to consider is regret. While most of us are afraid of making a bad decision that will negatively impact our lives or careers, it seems to me that when looking back over our lives to date, most of us are more upset by things we didn't do, not things we did do. We don't want to be nearing the end of our careers and wish we had been braver—to wish we had taken more bold action and achieved some stretch goal that may no longer be in reach.

When I interviewed Jason Ensher, I asked him if there was anything he might have done differently looking back over his career. His answer reflected this same sentiment:[8]

> *I might have approached things with a little more boldness. I might have taken the chance to start a business right out of graduate school, or right out of postdoc. Had I known more about the process, I might have said to my thesis or postdoc advisor, "There are some interesting things that we do here that might be valuable to other people." When you are younger and are used to working for very little money, it is easier to do this, and a university provides a fertile ground for good ideas. As you get older, a mortgage and other pressures begin to narrow your choices and make you a little less likely to take risks.*

I have certainly experienced the regret of not taking bolder actions at a few critical junctures in my early career. The occasion that stands out the most happened about 7 years into my industry career. SDL, the company in Silicon Valley where I'd begun my career was acquired by the much larger JDS Uniphase in 2001. As part of the inevitable consolidation that followed the bursting of the dot-com bubble later that same year, I accepted a mediocre position at the JDS site in Santa Rosa, California, to avoid being laid off when they closed the site where I was working. I had to move my family about 2 hours away, but at the time it seemed easier than finding a new job in the challenging job market that followed the bubble collapse. However, the reality of the job turned out to fall far below the image my wishful thinking had created, and six months later I found myself searching for a job again. Had I been braver during the previous transition and left the company to look for a position that fit me and my career goals, rather than taking the easy route, I likely would have saved a lot of energy and a move for my family.

The regret that I feel from that year in my life and career has motivated me to take bolder action in the years since. The most notable occasion is when I came to a transition point in my final year with my last employer. After a few years under the same management team, it had become clear to me that the company was not headed in the direction I'd been led to believe it was. I was faced with the choice of staying in a position that was clearly not going to give me the challenge and growth opportunities that I wanted, or leave and find something better. This was not an easy decision, as I had not spent any time searching for a new job and had no other prospects at the time. I decided to take a much bolder approach than I had previously and tendered my resignation. I decided that it was better to leave the job I knew was not going to work for me, rather than risk staying too long while I looked for a new job in my spare time.

During the newly acquired free time I had in the weeks after I left that job, I began looking more seriously at the private sector career speaking gigs I'd been doing as a volunteer side project for the previous 7 years. I'd collected many ideas for new seminars but hadn't had sufficient time and motivation to explore their full potential. Was there any chance I could actually build a business for myself out of all of these ideas?

I decided that I needed to pursue that chance, and I began building what ultimately became TurningScience the following year. Had I not taken the bold action to leave that job, I might have ended up 10 years later in my career still wondering if those ideas I'd jotted down on my iPad might actually be worth something. But that's not what happened. Looking back, it turned out to be absolutely the right decision.

If you aren't happy with the career you have now, or even if you are happy with it but dream of something more unique or more exciting or more rewarding, the private sector holds many options for you. The strengths of a scientist are valuable in so many different areas, and there are many different paths to take your career. Don't listen to anyone who tells you, 'That's not a good career path for a scientist.' Instead of thinking, 'maybe someday,' I suggest you think, 'there is no better time than right now.'

Be brave, be bold, and think big!

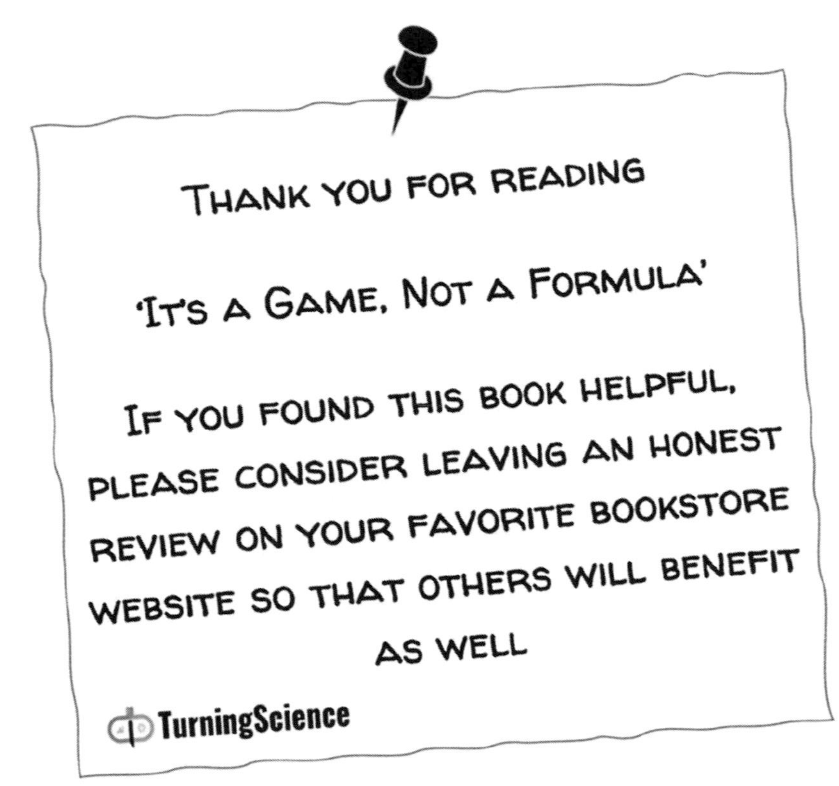
Thank you for reading
'It's a Game, Not a Formula'
If you found this book helpful, please consider leaving an honest review on your favorite bookstore website so that others will benefit as well
TurningScience

Interviewee Bios

Ashok Balakrishnan

Ashok Balakrishnan is the Chief Technology Officer and Co-CEO at Enablence Technologies in Kanata, Ontario. Ashok was a co-founder of Enablence Inc. and is a co-inventor of the key patents that define Enablence's proprietary technology. He served as their Vice President of Engineering for 16 years before his current role with the company, and he has extensive product development and commercialization expertise spanning several markets, including spectroscopy, telecommunications, and biophotonics. Prior to Enablence, Ashok was the manager of product integration at Optenia, a Staff Scientist at SDL (now JDS Uniphase), and a Photonics Development Scientist with Mitel Semiconductor (now Zarlink). He holds a PhD in Physics from the University of Toronto and was a postdoc researcher with the U.S. National Institute of Standards and Technology (NIST). In his spare time, Ashok enjoys running and spending time with his family.

Quotes and Interview excerpts: 15, 25, 42, 46, 97, 98

Tom Baur

Tom Baur is the founder, CTO, and chair of the board of Meadowlark Optics, Inc., a company founded in 1979 that designs, manufactures, and sells precision optics for polarization control. Prior to founding these companies, Tom was a research scientist at the National Center for Atmospheric Research (NCAR) in Boulder, Colorado. He has a BS in Astronomy from the University of Michigan and an MS in Astrophysics from the University of Colorado at Boulder.

Tom is the winner of the 2005 Bravo! Entrepreneur Award for Northern Colorado and a member of the Graduate School Advisory Council for the University of Colorado. In June of 2020, Tom established the first Endowed Chair at the Joint Institute of Laboratory Astrophysics (JILA), together with SPIE and the University of Colorado, Boulder. He lives on a ranch on the

Pawnee National Grassland in northeastern Colorado, where he and his wife, Jeanne, raise Black Angus cattle.

Quotes and Interview excerpts: 27, 28, 86–88, 104

Kate Bechtel

Kate Bechtel is a Biophotonics Fellow at Triple Ring Technologies in the San Francisco Bay area where she develops light-based instrumentation for the medical, biological, and environmental industries. She enjoys the opportunity her job gives her to experience the entire product development life cycle, from concept to fully qualified product.

Kate has a BS in Chemistry, Biochemistry, & Biophysics from Rensselaer Polytechnic Institute (RPI), a PhD in Physical & Analytical Chemistry from Stanford University, and did a postdoc in Biomedical Optics at the Massachusetts Institute of Technology (MIT). Following her postdoc, she transitioned to industry and accepted a Senior Scientist position at Triple Ring. In the 12 years since she joined, she has had a wide range of roles from technical development to business development to business area lead.

Her time in industry has given her valuable insight into where the strengths of a PhD scientist are particularly useful in a product development environment. She's also learned quite a bit about the patterns of thinking that many of us acquire in an academic setting that are not so helpful in an industry career.

Quotes and Interview excerpts: 16, 30, 31, 36, 50, 51, 131, 132, 137

Brit Berry–Pusey

Brit is a co-founder and the COO of Avenda Health, a medical device company developing a tissue ablation method for the treatment of prostate cancer in Santa Monica, California. She realized very early in her career that she wanted to use science to help people, and that desire led her to focus on the commercialization of biomedical technology. Now she helps lead a company that is aiming to improve the lives of millions of people while also striving to provide a rewarding career path for her employees.

Brit has a BS in Physics from the University of Utah and a PhD in Biomedical Physics from UCLA. After her PhD, she spent a couple of years at a mid-size medical device company and then a couple more years at a large multinational pharma company. In 2017, she left her job and founded Avenda Health with two contacts she met as a graduate student years earlier, demonstrating the value of making and maintaining a business network.

Brit places a high value on interacting with people who have more experience and requesting their help and advice. This practice has made a huge difference in her own life and career, leading her to opportunities that most early-career scientists never even consider, and setting her on a path toward improving the lives of millions of people around the globe, a goal that she traces back to her earliest childhood career plans.

Quotes and Interview excerpts: 61, 66, 91–93, 95, 111, 114, 125, 126, 140

Tanja Beshear

Tanja Beshear is a Senior Quality Systems Manager at Medtronic, leading the complaint handling activities for the Ventilation and Airways products. She received her MS in Physics from the University of Kaiserslautern in Germany, where she focused on experimental and applied physics, and completed her thesis in electron spectroscopy. She began her career developing fiber optic transceivers at Siemens in Berlin, where she quickly found a niche in Reliability Engineering and Failure Analysis. She continued her career in Quality Engineering at Siemens, Infineon, and Osram Opto-Semiconductors in San Jose, California.

Tanja also enjoyed working for start-ups and mid-size companies in Silicon Valley and later in Colorado, where she developed and led Reliability Engineering Programs. She transitioned into MedTech as a Quality Manager for Covidien, working on electro-surgical generators and instruments and led her team through the Medtronic acquisition. Tanja is a Senior Member of the American Society for Quality, and she is passionate about diversity, inclusion, and equity in the workplace, serving as a mentor and engaging in employee resource groups.

Quotes and Interview excerpts: 30, 95–97

Antoine Daridon

Antoine Daridon is a Business Development and Marketing Manager at Metrolab Technology SA in Geneva, Switzerland. Antoine holds a PhD in Analytical Chemistry from the Universität Bern in Switzerland; a Master's in Analytical Chemistry from the Université de Jussieu, Paris VI and the ESPCI, Paris Tech; and a BS in Organic Chemistry from the Université de Bretagne Occidentale.

After earning a PhD, Antoine took a postdoc position with the group of Nico de Rooij at the Université de Neuchâtel, where he developed microfluidic systems for environmental research. This experience led to an R&D position with Fluidigm in San Francisco, California, where he developed microfluidic assays for research in protein crystallography, cell-based assays, chemical synthesis, and genetic analysis. He has now acquired 15 years of program management experience and more than 20 years of R&D experience in the fields of biochemistry and microtechnology.

Quotes and Interview excerpts: xi, 30

Jason Ensher

Jason Ensher is Executive Vice President and Chief Technology Officer at Insight Photonic Solutions in Lafayette, Colorado. He earned a BS in Physics from the State University of New York at Buffalo. He earned a PhD in Physics from the University of Colorado at Boulder, where he worked with Dr. Eric Cornell, who won the 2001 Nobel Prize for demonstrating Bose–Einstein condensation. After completing his PhD, Jason went on to a postdoc at the University of Connecticut with Dr. Edward Eyler, and then began a career in industry. In addition to his current company, he has worked at ILX Lightwave, Precision Photonics Corporation, Ball Aerospace, and InPhase Technologies, all in the Boulder, Colorado area. In his spare time, Jason and his wife enjoy skiing, hiking, and swing dancing.

Quotes and Interview excerpts: 32, 48, 49, 58, 79, 83, 141

Peter S. Fiske

Peter S. Fiske is the Executive Director for the National Alliance for Water Innovation (NAWI) led by Lawrence Berkeley national laboratory in Berkeley, California. A native of Bethesda, Maryland, Peter has an AB from Princeton University, a PhD in Geological and Environmental Sciences from Stanford University, and an MBA from the University of California, Berkeley's Haas School of Business. In 1996, he was selected as a White House Fellow and served one year in the Pentagon as Special Assistant to the Under Secretary for Acquisition and Technology, Dr. Paul Kaminski. In 2001, Peter helped found RAPT Industries to commercialize a novel surface-processing technology developed at Lawrence Livermore Laboratories. In 2008, he took over the role of CEO at PAX Water Technologies, Inc. in San Rafael, California.

In addition to his day job, Peter is a nationally recognized lecturer on the subject of career development for scientists and engineers, and author of the book *Put Your Science to Work: The Take-Charge Career Guide for Scientists.* His columns have appeared in *Nature* and the online version of *Science.*

Quotes and Interview excerpts: 31, 33, 87, 88, 95–97, 100, 103, 105, 136

Sona Hosseini

Sona is a research and instrument scientist at the Jet Propulsion Lab (JPL) in Pasadena, California. She traces her interest in science back to a life-changing field trip to a NASA Center planetarium when she was seven years old. She has been enamored with astronomy ever since and decided to make astronomy her career.

Sona earned a BS in Physics from the University of Isfahan and an MS in Physics and Astronomy from the University of Zanjan, both in Iran. She also has an MS and PhD in Engineering Applied Science from the University of California, Davis. After completing a PhD, she took her current position at JPL where she leads a team developing spectroscopy instrumentation for space missions that will study the role of water in solar system formation.

I was excited to speak with Sona because, although she currently works in a national lab, she develops technology that is available for license to industry, and she has the unique opportunity in her job to interface with the public sector, the private sector, and academia. This gives her a unique view on the connections or the 'bridges' between these worlds. She also has some wonderful insights on personal growth and career planning that she has developed through some of her own career struggles.

Quotes and Interview excerpts: 28, 77, 115, 131, 135, 136

Marinna Madrid

Marinna is a physicist by training and co-founder of Cellino, a biotechnology company that is using lasers, robotics, and machine learning to build a fully automated approach to cell engineering. She and co-founder Nabiha Saklayen started Cellino in 2017, while Marinna was still a graduate student, after discovering that their research had valuable applications in synthetic biology.

Marinna has a BS in Biophysics from the University of California, Los Angeles, and a PhD in Applied Physics from Harvard University. She didn't discover her love for science until after she had graduated high school, but that has clearly not hindered her ability to make significant contributions in biophotonics. Beginning a startup was not what she had originally envisioned for her career, but after winning a startup pitch competition while still a student at Harvard, she and her research partner Nabiha decided to take the leap.

Marinna loves the variety and the endless opportunities to learn new skills that running a startup provides. She has a strong focus on communication and team dynamics as critical elements of success, and she brings this focus to her leadership role in Cellino.

Quotes and Interview excerpts: 44, 89, 90, 95, 96, 107–109

Roger McGowan

Roger McGowan is a Sr. Research Fellow at Boston Scientific Corporation (BSC) in Maple Grove, Minnesota. He received a PhD in Physics from Colorado State University, where he performed experiments in light forces on atoms with Dr. Siu Au Lee. After receiving his PhD, Roger accepted a postdoctoral position with Dr. Dan Grishkowski at Oklahoma State University. He then returned to the Minneapolis/St. Paul area where he grew up, working at Imation and ADC Telecommunications before discovering the joy of the medical device industry.

For the past eighteen years he has been a technical leader across franchises for BSC's new product development. A key focus for him has been to drive deep technical understanding into the process, design, and robustness of BSC's products. This led to the opportunity for Roger to lead small teams working closely with external physician partnerships (Mayo, Emory, Mass General Brigham, Stanford). These teams utilize rapid prototyping combined with frequent physician evaluation to accelerate device development cycles to First in Human trials. In 2019, Roger received the highest technical achievement award within Boston Scientific for these leadership, business, mentorship, and career accomplishments. Roger has been awarded 14 patents and has an additional 17 patents pending in medical device design.

In high school, Roger was a self-proclaimed "gear head" and says he gained his interest in mechanical design and operation while working on his 1970 Dodge Challenger. This activity influenced his professional life and his decision to pursue a technical career. He also credits his extensive work with youth activities in college with helping him understand how to work with people, a skill that he has found very useful in his career.

Quotes and Interview excerpts: 57, 115, 122

Chris Myatt

Chris Myatt is the founder and CEO of LightDeck Diagnostics, a medical diagnostics company that he founded as mBio Diagnostics in 2009 in Boulder, Colorado. mBio Diagnostics was spun out of Precision Photonics Corporation (PPC), a company he founded with his wife, Sally Hatcher, in 2000 to develop precision optical components, coatings, and assemblies for the telecommunications, aerospace and defense, biomedical, and semiconductor industries.

Chris earned a BS in Physics and a BA in Mathematics at Southern Methodist University, and a PhD in Atomic Physics at the University of Colorado. As a graduate student, he worked with Carl Weiman on the Bose–Einstein Condensation (BEC) project for which Carl received the Nobel Prize in physics in 2001. After graduate school, he went on to a postdoctoral fellowship at NIST in Boulder working on ion trapping and foundational quantum computing experiments under the direction of Dave Wineland. In his spare time Chris enjoys cycling and hockey.

Quotes and Interview excerpts: 2, 44, 81, 83, 86, 90, 91, 99, 106, 108

Kerstin Schierle–Arndt

Kerstin Schierle–Arndt is a PhD chemist with a specialization in electrochemistry. She completed all of her university studies at the University of Bonn in Germany and then began her career at BASF in Ludwigshafen, Germany where she still works today as a Vice President of Research Inorganic Materials and Synthesis.

Kerstin knew early on that she wanted to do chemistry on a large scale, so a chemical company like BASF was a good fit. She began in research, but then took the opportunity to try different roles that exposed her to the other aspects of the business before returning to research. She feels that this experience has made her a much better research manager, because she now has a better understanding of how to ensure the research projects her team pursues will make money for the company. The same applies for the selection of research topics.

Kerstin believes that enjoying the work she does is the best approach to work–life balance, and work she finds rewarding is more important to her than pursuing any particular status or job title. With her job at BASF, she has been able to achieve this and still have a rewarding life outside of her work.

Quotes and Interview excerpts: 37, 38, 118

Scott Sternberg

Scott Sternberg is the Executive Director of the Boulder Economic Council and the Associate Vice President for Economic Vitality at the Boulder Chamber, both in Boulder, Colorado. When I interviewed him for this book, he was the Executive VP of Services at Vaisala, a global environmental and industrial monitoring device company headquartered in Finland. He later went on to become President of Vaisala Inc., the company's U.S. subsidiary headquartered in Louisville, CO.

Prior to Vaisala, Scott held many roles for Roper Industries/Photometrics in Tucson, AZ. He has a BS in Physics from SUNY College at Cortland and an MS in Physics from Colorado State University.

Quotes and Interview excerpts: 62, 76, 138

Yasaman Soudagar

Yasaman is a co-founder and the CEO of Neurescence, a medical device company building brain-imaging tools in Toronto, Ontario. Their fluorescence microscope provides functional imaging of the brain and spinal cord, helping researchers better understand brain diseases such as Alzheimer's and develop effective therapeutics.

Yasaman has a PhD in Physics from the École Polytechnique de Montréal, with a specialization in experimental quantum optics. Although she originally planned a career in academia, she ended up transitioning into industry right after her PhD. In June of 2014, she started Neurescence based on an opportunity she realized while developing laser systems for surgical applications.

Although running a startup is really hard work, Yasaman says she doesn't really feel like she is working, because it is so fulfilling and has given her countless opportunities for personal growth. She credits her success to her science PhD training, a strong network, and the willingness to jump in and figure it out as she goes, something that did not originally come so naturally to her.

Quotes and Interview excerpts: 20, 45, 46, 61, 64, 67–70, 93, 94, 96, 99, 104, 110, 139, 140

Christina C.C. Willis

Christina is a laser scientist and the 2019–2020 OSA/SPIE Arthur H. Guenther Congressional Fellow through the AAAS Science and Technology Policy Fellowship program. She is serving her year in the United States Senate in Washington, D.C. She has a BA in Physics from Wellesley College and an MS and PhD in Optics from the College of Optics and Photonics (CREOL) at the University of Central Florida. After graduate school, she spent four years in the private sector, working as a laser scientist with Vision Engineering Solutions and Fibertek, Inc.

Christina believes in the value of networking and has invested significant energy in developing her networking skills. This has resulted in several valuable opportunities, including a position on the Board of Directors with SPIE and an invitation to write a book on networking. Her book, *Sustainable Networking for Scientists and Engineers*, was published by SPIE Press in early 2020. She is also a writer of fiction and is currently querying her first science fiction novel.

Christina has also learned the value of persistence when facing challenging problems and embracing uncomfortable situations as an opportunity to learn and grow. These strengths have resulted in some unique experiences that I found very valuable to learn about. I hope you find her interview as valuable as I did.

Quotes and Interview excerpts: 29, 42, 116, 117, 121, 127, 128, 133, 134

Oliver Wueseke

Oliver (Olli) Wueseke is the founder and CEO of Impulse Science, a science consultancy that guides biotech startup teams in research strategy and product development. Before starting Impulse Science, Olli was the Director of R&D at the startup System1 Biosciences in the San Francisco Bay Area in California.

Olli has a BS in Biology and a PhD in Molecular Biology from the Max Planck Institute for Molecular Cell Biology and Genetics in Dresden,

Germany. Following his PhD, he did a postdoc at the Institute for Molecular Biotechnology in Vienna, Austria.

The strengths that Olli has found most valuable in his career so far include empathy and the ability to inspire others. These are perhaps not traditional strengths for a scientist, but he has found they make him a great leader of product development teams. In early 2020, he decided to leverage these strengths fully. He started his own company, helping startups to successfully navigate the challenges of bringing a new product to market.

Quotes and Interview excerpts: 3, 4, 18, 27, 74, 82, 117, 118, 122–130

Notes

Preface

1. D. M. Giltner, "Improving the STEM PhD's transition into a private sector career," *Proc. SPIE* **11480**, 1148007 (2020).
2. D. M. Giltner, *Turning Science into Things People Need*, Wise Media Group, Denver, CO, 86 (2010).
3. D. M. Giltner and P. Gramlich, "How to get research funding from industry," *Chemistry World* (October 30, 2019).

Chapter 1: Introduction

1. D. M. Giltner, *Turning Science into Things People Need*, Wise Media Group, Denver, CO, 21- 29 (2010).
2. Interview with Oliver Wueseke, founder and CEO of Impulse Science, (May 2020)
3. D. Knuth, "The Computer as Mastermind", *J. Recr. Math.*, 9, 1976-77
4. P. A. Offit, Vaccinated: One Man's Quest to Defeat the World's Deadliest Diseases, Harper Collins, 11 (2007).
5. U.S. Center for Disease Control Measles Fact Sheet, https://www.cdc.gov/globalhealth/ measles/resources/measles-rubella-initiative-fact-sheet.html.
6. S. Combs, *This virologist saved millions of children—and stopped a pandemic*, National Geographic, May 29, 2020, https://www.nationalgeographic.com/history/article/virologist-maurice-hilleman-saved-millions-children-stopped-pandemic
7. J. C. Simpson, *The Man Who Beat the 1957 Flu Pandemic*, Scientific American, April 19, 2020, https://blogs.scientificamerican.com/observations/the-man-who-beat-the-1957-flu-pandemic/
8. P. A. Offit, *Hilleman: A Perilous Quest to Save the World's Children*, Medical History Pictures, Inc. (2016).

Chapter 2: Rules of the Game

1. D. M. Giltner, *Turning Science into Things People Need*, Wise Media Group, Denver, CO, 77 (2010).
2. Interview with Kate Bechtel, Biophotonics Fellow at Triple Ring Technologies, Interview, May 2020
3. From the text in Alfred Nobel's will, https://www.nobelprize.org/alfred-nobel/full-text-of-alfred-nobels-will-2.
4. C. Bartneck and M. Rauterberg, "Physics Nobels should favour inventions," *Nature* **448**(7154), 644 (2007) [doi: 10.1038/448644c].
5. C. Bartneck and M. Rauterberg, "The asymmetry between discoveries and inventions in the Nobel Prize in Physics," *Technoetic Arts* **6**(1) (2008) [doi: 10.1386/tear.6.1.73/1].
6. From Scott's pitch delivered at the 1 Million Cups startup pitch mentoring event in Boulder, Colorado on May 23, 2018.
7. Interview with Oliver Wueseke, founder and CEO of Impulse Science, (May 2020)
8. Interview with Yasaman Soudagar, co-founder and CEO of Neurescence in Toronto, Ontario (May 2020).
9. I'm quite certain that any scientist who has experience securing funding and navigating university administration bureaucracy will state more emphatically than I have here that academic research is indeed a very competitive game. I agree completely. In this book I emphasize the differences between academia and industry in order to grab the attention of those who entered the private sector with the view or academia they had as PhDs and postdocs.

Chapter 3: The PhD Stereotypes

1. A. M. Porter, Physics PhDs Ten Years Later: Success Factors and Barriers in Career Paths, American Institute of Physics (2019).
2. D. M. Giltner, *Turning Science into Things People Need*, Wise Media Group, Denver, CO, 79 (2010).
3. D. M. Giltner, *Turning Science into Things People Nee*d, Wise Media Group, Denver, CO, 52 (2010).
4. R. Czjuko and G. Anderson, *Common Careers of Physicists in the Private Sector*, American Institute of Physics (2015).
5. D. M. Giltner, *Turning Science into Things People Need*, Wise Media Group, Denver, CO, (2010).

6. Interview with Oliver Wueseke, founder and CEO of Impulse Science (May 2020).
7. Interview with Sona Hosseini, research and instrument scientist at the Jet Propulsion Lab (June 2020).
8. Interview with Christina C. C. Willis, laser scientist and the 2019-2020 OSA / SPIE Arthur H. Guenther Congressional Fellow through the AAAS Science and Technology Policy Fellowship program (April 2020).
9. Interview with Kate Bechtel, Biophotonics Fellow at Triple Ring Technologies (May 2020).
10. D. M. Giltner, *Turning Science into Things People Need*, Wise Media Group, Denver, CO, 85 (2010).
11. D. M. Giltner, *Turning Science into Things People Need*, Wise Media Group, Denver, CO, 35 (2010).
12. D. M. Giltner, *Turning Science into Things People Need*, Wise Media Group, Denver, CO, 46 (2010).
13. D. M. Giltner, *Turning Science into Things People Need*, Wise Media Group, Denvcr, CO, 61 (2010).
14. M. Shepherd, 9 Tips for Communicating Science to People Who Are Not Scientists, Forbes (November 22, 2016).
15. R. Olson, Houston, We Have a Narrative: Why Science Needs a Story, University of Chicago Press (2015).
16. Interview with Kerstin Schierle-Arndt, Vice President of Research Inorganic Materials and Synthesis with BASF (August 2020)

Chapter 4: Your Private-Sector Playbook

1. In my work, I've found two different types of academic research groups that tend to encourage behaviors and working habits that are helpful in the private sector. The first type is a group that has built successful industry collaborations. In working closely with companies, these groups have learned to move quickly and focus on work that has value in industry. The second type is a research group who views themselves in a competition to be the first to demonstrate something new. For this second type of example, see my interviews with Jason Ensher and Chris Myatt in my book *Turning Science into Things People Need.*
2. Interview with Christina C. C. Willis, laser scientist and the 2019-2020 OSA / SPIE Arthur H. Guenther Congressional Fellow through the AAAS Science and Technology Policy Fellowship program (April 2020).

3. C. R. Cook and J. C. Graser, *Military Airframe Acquisition Costs: The Effects of Lean Manufacturing,* Santa Monica, CA: RAND Corporation, pg 103, www.rand.org/pubs/ monograph_reports/MR1325.html (2001).
4. E. M. Kaitz, Overhead Costs and Rates in the U.S. Defense Industrial Base. Volume 1, pg 35-36 (1980).
5. D. M. Giltner, *Turning Science into Things People Need*, Wise Media Group, Denver, CO, 78 (2010).
6. Interview with Marinna Madrid, co-founder of Cellino Technologies (May 2020).
7. D. M. Giltner, *Turning Science into Things People Need*, Wise Media Group, Denver, CO, 24 (2010).
8. Interview with Yasaman Soudagar, co-founder and CEO of Neurescence in Toronto, Ontario (May 2020).
9. I'm a fan of the 5 × 5 risk matrix that is based on identifying the probability and the severity of a risk on a 1–5 scale. The goal of this activity is to prioritize all of the imaginable risks into one of three categories: High – risks that need to be addressed immediately; medium – risks that should be addressed before the project is over, but not right away; and low – risks that can be ignored.
10. D. M. Giltner, *Turning Science into Things People Need*, Wise Media Group, Denver, CO, 57- 64 (2010).
11. Interview with Scott Kelly on *The Tim Ferris Show*, November 7, 2020, https://tim.blog/2020/11/07/scott-kelly-transcript/
12. Interview with Kate Bechtel, Biophotonics Fellow at Triple Ring Technologies (May 2020).
13. R. Koch, The 80/20 Principle: The Secret of Achieving More with Less, Doubleday, New York, (1998)
14. S. R. Covey, *The Seven Habits of Highly Effective People: Restoring the Character Ethic*. Simon and Schuster, New York (1989). The matrix is found in the chapter on Habit 3: Put First Things First. This tool is also sometimes referred to as the Eisenhower matrix, in reference to the U.S. president Dwight Eisenhower, who was known for not letting urgency dictate what he considered important.
15. Tim Ferriss' blog post on October 25, 2007, *The Art of Letting Bad Things Happen (and Weapons of Mass Distraction)*, https://tim.blog/2007/10/25/weapons-of-mass-distractions-and-the-art-of-letting-bad-things-happen.
16. D. M. Giltner, *Turning Science into Things People Need,* Wise Media Group, Denver, CO, 18 (2010).
17. S. Godin, *Linchpin: Are You Indispensable?*, Penguin Group (USA) Inc., New York 80 (2010).

18. S. Godin, *Linchpin: Are You Indispensable?*, Penguin Group (USA) Inc., New York, 57 (2010).
19. Tom Petty in the song "Crawling Back to You," from the *Wildflowers* album, Warner Bros. (1994).
20. Interview with Brit Berry–Pusey, co-founder and COO of Avenda Health (May 2020).
21. EQ stands for emotional quotient and is a term used to signify emotional intelligence, which describes the capacity of an individual to be aware of, control, and express one's emotions, and to handle interpersonal relationships judiciously and empathetically.
22. Interview with Scott Sternberg, Executive Director of the Boulder Economic Council (March 2010)
23. R. B. Cialdini, *Influence: Science and Practice*, Harper Collins College Publishers, New York (1993).
24. P. A. Offit, Hilleman: A Perilous Quest to Save the World's Children, Medical History Pictures, Inc. (2016).
25. D. Golcman, Emotional Intelligence: Why It Can Matter More Than IQ, Bantam Books, New York (1995).
26. In addition to Goleman's book listed above, here are two more useful references:
    - D. H. Pink, Drive: The Surprising Truth About What Motivates Us, Riverhead Books, New York (2009).
    - N. Morgan, R. B Cialdini, L. A. Hill, and N. Duarte, *Influence and Persuasion (HBR Emotional Intelligence Series),* Harvard Business Review Press (2017).

Chapter 5: The R&D Mindsets

1. J. A. Shaw, *Synergistic development of optics education and industry in a small university town*, Proc. SPIE 9793, Education and Training in Optics and Photonics: ETOP 2015, 97932J (8 October 2015); doi: 10.1117/12.2223218
2. Interview with Oliver Wueseke, founder and CEO of Impulse Science (May 2020).
3. Interview with Scott Sternberg, Executive Director of the Boulder Economic Council (March 2010)
4. Interview with Sona Hosseini, research and instrument scientist at the Jet Propulsion Lab (June 2020).

5. The 2001 Nobel Prize in Physics was awarded jointly to Eric A. Cornell, Wolfgang Ketterle, and Carl E. Wieman "for the achievement of Bose–Einstein condensation in dilute gases of alkali atoms, and for early fundamental studies of the properties of the condensates." https://www.nobelprize.org/prizes/physics/2001/summary.
6. D. M. Giltner, *Turning Science into Things People Need*, Wise Media Group, Denver, CO, 24 (2010).
7. D. M. Giltner, *Turning Science into Things People Need*, Wise Media Group, Denver, CO, 60 (2010).

Chapter 6: Startups – The Ultimate Game!

1. There are many books and articles on how to be a successful entrepreneur. Like advice on playing a game, opinions vary on the quality of each reference. Rather than try to give you some that I promise will make you successful, I'll give you some of my own personal favorites. Each gives a unique and valuable perspective:
   - P. A. Thiel and B. Masters, Zero to One: Notes on Startups, or How to Build the Future, Westminister, MD, 2014
   - B. Horowitz, The Hard Thing About Hard Things: Building a Business When There Are No Easy Answers, Harper Business, 2014
   - J. Fried and D. Heinemeier Hansson, *Rework*, Crown Business, 2010
   - C. M. Christensen, The Innovator's Dilemma: When New Technologies Cause Great Firms to Fail, Boston, MA: Harvard Business School Press, 1997
   - M. Chang, Toward Entrepreneurship: Establishing a Successful Technology Business, Milton Chang, 2011
   - J. Haden, "8 Qualities of Fearless Entrepreneurs," Inc. October 8, 2012, https://www.inc.com/jeff-haden/8-qualities-of-fearless-entrepreneurs.html
2. D. M. Giltner, *Turning Science into Things People Need*, Wise Media Group, Denver, CO, 21-29 (2010).
3. D. M. Giltner, *Turning Science into Things People Need*, Wise Media Group, Denver, CO, 49-56 (2010).
4. Press Release, Meadowlark Optics: https://www.meadowlark.com/store/PDFs/Meadowlark-Optics-FOR-IMMEDIATE-RELEASE-$2.5M-Endowment.pdf
5. D. M. Giltner, *Turning Science into Things People Need*, Wise Media Group, Denver, CO, 41-48 (2010).

6. Interview with Marinna Madrid, co-founder of Cellino Technologies (May 2020).
7. Cellino website: http://cellinobio.com
8. LightDeck Diagnostics company website: https://lightdeckdx.com/
9. J. Shieber, LightDeck Diagnostics has $11 million for its new, high-speed way to test for COVID-19, sepsis and heart attack, TechCrunch, October 29, 2020 https://tcrn.ch/34BxoWg
10. Interview with Brit Berry–Pusey, co-founder of Avenda Health (May 2020).
11. Interview with Yasaman Soudagar, co-founder and CEO of Neurescence in Toronto, Ontario (May 2020).
12. This book doesn't present much about how to network, but it is a critical skill for success in the private sector game. The following book is a wonderful reference for scientists: C. C. C. Willis, *Sustainable Networking for Scientists and Engineer*s, SPIE–The International Society for Optical Engineering (2020).
13. Giltner, D. M, *Turning Science into Things People Need*, Wise Media Group, Denver, CO, 31-39 (2010).
14. Giltner, D. M, *Turning Science into Things People Need*, Wise Media Group, Denver, CO, 75-81 (2010).
15. Interview with Chris Myatt, founder and CEO of LightDeck Diagnostics (then mBio Diagnostics), February 2010.
16. Moore, Geoffrey A. Crossing the Chasm: Marketing and Selling Technology Products to Mainstream Customers. New York, N.Y.: Harper Business, 1991. Print.
17. P. Fiske, "Life as an Entrepreneur Lecture Series," University of California Berkeley, October 18, 2007, https://www.youtube.com/watch?v=6yEj-yfOKJk
18. E. Ries, The Lean Startup: How Today's Entrepreneurs Use Continuous Innovation to Create Radically Successful Businesses, New York: Crown Business, 2011
19. The MVP concept was first coined in 2001 by management consultant Frank Robinson (http://www.syncdev.com) and is now a common term in the startup world. Eric Ries comes from a software background where the ease of sending software updates to customers makes the MVP concept particularly straightforward. Customers who receive the MVP can easily be sent updates as the product is further developed. This is not so straightforward with hardware products, so the approach needs to be modified. Either the MVP needs to be complete enough that the early-adopter customers have a usable product for the long-term, or the

company needs to consider a hardware-upgrade option for those customers who are willing to test the MVP and give useful feedback.

20. Carl Weiman is an American physicist who produced the first Bose–Einstein condensate, working with Eric Cornell at the University of Colorado Boulder in 1995, for which they were awarded the Nobel Prize in Physics in 2001 along with Wolfgang Ketterle. Carl was Chris Myatt's PhD advisor.
21. C. S. Dweck, *Mindset: The New Psychology of Success.* New York: Ballantine Books, 2008.
22. E. Meceda, 'Mindset Course,' https://www.mindset-course.com/
23. B. Horowitz, The Hard Thing About Hard Things: Building a Business When There Are No Easy Answers, Harper Business, 2014

Chapter 7: Your Career Is a Game

1. S. Sinek, Start with Why: How Great Leaders Inspire Everyone to Take Action, Penguin Books, 2011
2. Interview with Brit Berry–Pusey, co-founder and COO of Avenda Health (May 2020).
3. Interview with Sona Hosseini, research and instrument scientist at the Jet Propulsion Lab (June 2020).
4. D. M. Giltner, *Turning Science into Things People Need,* Wise Media Group, Denver, CO, 11-19 (2010).
5. Interview with Christina C. C. Willis, laser scientist and the 2019–2020 OSA / SPIE Arthur H. Guenther Congressional Fellow through the AAAS Science and Technology Policy Fellowship program (April 2020).
6. Interview with Oliver Wueseke, founder and CEO of Impulse Science (May 2020).
7. Interview with Kerstin Schierle–Arndt, Vice President of Research Inorganic Materials and Synthesis with BASF (August 2020)
8. D. M. Giltner, *Turning Science into Things People Need,* Wise Media Group, Denver, CO, 57-64 (2010).
9. Quote attributed to Branch Rickey, American baseball player and sports executive. Rickey was instrumental in breaking Major League Baseball's color barrier by signing player Jackie Robinson.
10. Bill Burnett and Dave Evans, *Designing your Life*, Alfred A. Knopf, New York, NY, xxi (2019)
11. Graphic designer and educator Debbie Millman guides her students through an approach for visualizing the life they desire with an exercise

she calls "Your Ten-Year Plan for a Remarkable Life." I performed this exercise in 2017 when I was considering a career change, and it was instrumental in my decision to leave my job and start TurningScience. This experience was transformative for me, and I highly recommend it for everyone. Debbie outlines this approach in a great interview on the Tim Ferriss Show in January 2017. The interview can be found here: https://tim.blog/2017/01/12/how-to-design-a-life-debbie-millman/

12. Bill Burnett and Dave Evans, *Designing your Life*, Alfred A. Knopf, New York, NY, 113 (2019)
13. Interview with Kate Bechtel, Biophotonics Fellow at Triple Ring Technologies (May 2020).
14. D. M. Giltner, *Turning Science into Things People Need,* Wise Media Group, Denver, CO, 41-48 (2010).
15. Interview with Scott Sternberg, Executive Director of the Boulder Economic Council (March 2010)
16. Interview with Yasaman Soudagar, co-founder and CEO of Neurescence in Toronto, Ontario (May 2020).

About David M. Giltner, PhD

David Giltner is an internationally recognized author, speaker, and mentor on the topics of private sector career development and technology commercialization. He began his science career with the intention of following the 'traditional' path to become a tenured research professor. However, during the final year of his PhD work, he decided that the academic career path was not for him after all. He completed his PhD and found his first job with SDL Inc in San Jose, California, developing scientific instruments based on semiconductor laser technology. Since then, he has spent more than two decades commercializing cutting-edge photonics technologies for a variety of applications including optical communications, remote sensing, and scientific instrumentation.

After beginning his career in the private sector, David quickly moved out of hands-on lab work into roles leading product development teams, and then into customer-facing roles in product management and business development. His diverse career path has included working with companies ranging from very large to very small, and this has given him a broad perspective on the private sector and what is required to turn a new technology into a successful commercial product. The extensive time he has spent working with both technical teams and company executives has taught him to function as an 'interpreter' between the different languages and cultures of MBAs and PhDs.

David has a BS in Physics from Truman State University and a PhD in Physics from Colorado State University. In 2010, he published *Turning Science into Things People Need* and began speaking to early-career scientists interested in building private sector careers. In 2017 he started TurningScience to provide training and support for scientists of all disciplines seeking to enter the private sector as employees, academic collaborators, or entrepreneurs. In addition to giving career-based seminars and workshops for scientists, he works as a consultant and mentor, for entrepreneurs and technology companies in the U.S. and Europe. He has a passion for travel, food, music, and the outdoors; and can usually be found attempting to combine these interests either abroad or in his home state of Colorado in the United States.